Performance of Public and Private Mining Firms in India

Performance of Public and Private Mining Firms in India: In Productivity, Environmental and Social Dimensions

By

Amarendra Das

CAMBRIDGE SCHOLARS

PUBLISHING

Performance of Public and Private Mining Firms in India:
In Productivity, Environmental and Social Dimensions,
by Amarendra Das

This book first published 2013

Cambridge Scholars Publishing

12 Back Chapman Street, Newcastle upon Tyne, NE6 2XX, UK

British Library Cataloguing in Publication Data
A catalogue record for this book is available from the British Library

ISBN (10): 1-4438-5243-0, ISBN (13): 978-1-4438-5243-2

To the people of mining regions…

CONTENTS

LIST OF TABLES

FOREWORD

The extraction and use of natural resources in developing countries are undergoing a notable transition. First, there has been an unprecedented increase in the demand for such resources driven by the economic growth of emerging economies in Asia, Latin America and Africa. This growth in demand creates pressure to increase the supply of natural resources. Secondly, there have been different changes in the ownership of these resources or the ownership of firms involved in their extraction. On the one hand, there are attempts to assert national ownership and nationalization in certain countries where foreign firms had a near monopoly over these resources due to historical reasons or policies of past regimes. On the other hand, there is a growing pressure for privatisation where only the government firms were extracting the resources, partly driven by the inability of these firms to meet the growing demand in an efficient manner. Though the extraction of natural resources had always caused negative externalities in terms of relocation of settlements and environmental pollution, people in different parts of the world are currently asserting their rights to be free from such externalities.

These are the issues analyzed in the doctoral thesis of Amarendra Das, which is now being published as this book. The most important contribution is the systematic comparison of the performance of private and public-sector mining firms in terms of productivity, compliance of environmental regulations and mitigation of social costs. Though there can be multiple theoretical propositions about the better performance of these two types of firms (private or public sector), ex-ante theoretical prediction of the superiority of any one type is difficult, and this issue needs to be analysed empirically in the specific socio-economic context. Hence the insights of the analysis reported in this book are important for the current situation in India.

The depth of analysis is determined by the availability of data, and it is not easy to have or collect exhaustive data which can be used to compare different firms in terms of environmental pollution or the social costs. Hence the chapters analysing these issues are more indicative in nature. However, it was relatively easy to have comprehensive and comparable data on productivity of the firms. Hence, the insight of the comparative analysis of productivity provides a very reliable picture of the situation.

In summary, the thesis and the book are very important since they analyse certain issues which are very relevant in the contemporary context in a systematic manner.

V. Santhakumar

PREFACE

The mining industry across the world has drawn the attention of the public and researchers for many reasons. Starting from the Mid 1980s, many countries across the globe have opened up their mining industry for private sector participation with a view to receiving more investment and better technology, which will result in augmenting the overall output and productivity of the industry. At the same time, private participation in the mining industry has raised concerns over environmental degradation, loss of livelihood and involuntary displacements without proper compensation. Such concerns have not only been raised in developing countries, but also in economically advanced countries. For example, in fear of severe environmental damage and loss of livelihood, the indigenous (adivasi) people living adjacent to the Niyamagiri hill in the Koraput district of Odisha, India, opposed the mining and aluminium project of the Vedanta Aluminium Company. The mass protest in India drew global attention, and consequently the Church of England withdrew its' share invested in the Vedanta Aluminium Company.

Against this backdrop, I undertook a comprehensive study to examine the performances of public and private sector mining firms in three dimensions: productivity, environmental compliance and social compliance. The word social compliance has been defined as the compensation provided by the mining firms for direct and indirect losses caused to the land surrendering households as per the promises made during the land acquisition. The results of the studies were submitted as a dissertation for my PhD degree at the Centre for Development Studies, Thiruvananthapuram, Kerala, which is affiliated to the Jawaharlal Nehru University, New Delhi, India.

I am extremely grateful to Cambridge Scholars Publishing, UK, for showing interest to publish my research work. The completion of this research work involved a great deal of help from a number of individuals in a direct and indirect manner. Therefore, upon the completion of this work, I feel it is my sincere duty to put on record the generous support I have received from all of them. I am sure that my words will be scant to appreciate the contributions of every individual. First of all, I express my deep sense of gratitude to my PhD supervisors –Dr Vellapan Santhakumar and Dr Mavanoor Parameswaran— for their continuous cooperation and moral support in completing this research work. The Global Development

Award, 2008, for my paper entitled '*Do Firm Ownership and Competition Have Bearing on Productivity? An Enquiry of Indian Mining Industry from 1988-89 to 2005-06*', which came from Chapter Three of this book would not have been possible without their meticulous guidance.

While gathering the primary and secondary data a number of officials were helpful at the Indian Bureau of Mines, Nagpur, Bhubaneswar; State Pollution Control Board, Odisha; Directorate of Mines, Bhubaneswar, Department of Steel and Mines; Revenue Department, Government of India, collectorates of Jajpur, Angul, Kendujhar; Officers of Coal India limited at Mahanadi Coalfield limited, Jagannath area, deputy director of mines, Jajpur and Kendujhar; Mines officer, Kendujhar and Joda; Ministry of Environment and Forest, Bhubaneswar Branch; General Manager of Orissa Mines Corporation (OMC) and the Mines officer, OMC at SGBK mines, Guruda; Deputy General Manager of Daitari Mines, Kaliapani; Mines officer of Orissa Stewardess Limited (OSL) at Joda; Mines officer of TISCO at Sukinda; Director and Social Welfare officer of Industrial Development Corporation, Odisha at JK Road. Mr. R. N Sahoo, the director of mines Odisha lent his generous support to provide me with the data and logistic assistance to carry out the survey in remote mining areas. I am thankful to all of them. The primary survey conducted for examining the social performance of mining firms would not have been possible but for the cooperation and contribution from Santa, Ratnakar, Ajay, Braja and the respondents of surveyed villages – Guruda, Balada and Palsa, of Jajpur district and Gurujang, Ostapal and Tailangi of Kendujhar district. I express my sincere gratitude to all of them. I am grateful to the entire family of Aswini Bhai for providing me their unconditional love and support during the research consultations and data collection in Bhubaneswar.

I am extremely grateful to Mr. Anil Menon for carefully editing the manuscript in due time. I would also like to thank William, Braja, Priyajit, Atish, Sanjaya, for their enriching discussions on my work at different points in time. At home, Jepa, Bapa, Maa, Bibhunandini, Nana, Bhauja, Santa and Neta always remain as my perennial source of love and inspiration. My sincere gratitude to all of them.

Amarendra

ABBREVIATIONS AND ACRONYMS

CEC	Central Empowered Committee
CIL	Coal India Limited
CMA	Compania Minera Antamina
CMIE	Centre for Monitoring Indian Economy
CSE	Centre for Science and Environment
CSO	Central Statistical Organisation
CSR	Corporate Social Responsibility
EIA	Environmental Impact Assessment
EMP	Environmental Management Plan
ENAMI	Empresa Nacional de Mineria
FDI	Foreign Direct Investment
FIMI	Federation of Indian Mining Industries
FIPB	Foreign Investment Promotion Board
GDP	Gross Domestic Product
GFA	Gross Fixed Asset
GMM	Generalised Method of Moments
GoI	Government of India
GSI	Geological Survey of India
HDI	Human Development Index
IDC	Industrial Development Corporation
IPR	Industrial Policy Resolution
LAA	Land Acquisition Act
M&Q	Mining and Quarrying
MCDR	Mineral Concession and Development Rules
MCR	Mineral Concession Rules
MEDI	Multidimensional Environmental Defiance Index
ML	Mining Lease
MMRD	Mines and Minerals Regulation and Development
MoC	Ministry of Coal
MoEF	Ministry of Environment and Forest
NELP	New Exploration Licensing Policy
NIC	National Industrial Classification
NMP	National Mineral Policy
NSDP	Net State Domestic Product
OIL	Oil India Limited
OLS	Ordinary Least Squares

OMC	Orissa Mining Corporation
ONGC	Oil and Natural Gas Corporation
OP	Olley and Pakes
PAP	Project Affected Persons
PE	Private Enterprises
PL	Prospecting License
POSCO	Pohang Steel Company
PSE	Public Sector Enterprises
PSU	Public Sector Undertakings
RBI	Reserve Bank of India
RP	Reconnaissance Permit
RSI	Relationship Specific Investment
SCCL	Singareni Collieries Company Limited
SoE	State Owned Enterprises
SPCB	State Pollution Control Board
SPM	Semi Parametric Method
SPM	Suspended Particulate Matter
TFP	Total Factor Productivity
TRR	Total Revenue Receipts
UN	United Nations
USA	United States of America
VAT	Value Added Tax
VCHC	Value of Capital at Historical Cost
VCRC	Value of Capital at Replacement Cost
WWF	World Wildlife Fund

CHAPTER ONE

INTRODUCTION

"Mines are the source of wealth; from wealth comes the power of government..."
—Kautilya, as translated in Rangarajan, 1992; p.84

1.1 Motivation

Minerals are considered a national treasure endowed by nature, and obtain enormous value for being finite and non-renewable. They constitute the vital raw materials for many basic industries and are a major resource for development (GoI, 2010). The mining industry provides the basic inputs to a number of key sectors of the economy like power and manufacturing. Thermal power generation, for example, depends upon the uninterrupted supply of coal; production of automobiles depends upon the supply of aluminium and steel and construction activities depend upon the supply of steel and cement. The growth of the manufacturing sector is thus largely influenced by the growth of mining industry. Low production and productivity in the mining industry may constrain the production and productivity in the related manufacturing sectors and dampen overall economic growth. In order to overcome these bottlenecks, from the mid-1980s many developing countries have been diluting the monopoly of the public sector on the extraction of non-renewable natural resources and attracting private players (both domestic and foreign) to take part in exploration and extraction. Several countries have been amending their mining legislation and putting in place the necessary institutions for this purpose. By 1993, around 70 countries in Latin America, Africa and Asia-Pacific had fully liberalized their mining laws and implemented deregulation in a wide range of areas, including land rights, mineral rights, taxation and environmental protection, in order to attract foreign mining investors (TWN, 1997).

The participation of private players is expected to increase production and overall productivity[1] of the mining sector through direct and indirect effects. Under direct effects, it is presumed that private firms are more efficient than their public counterparts and their participation will bring in more capital, better technology and superior managerial skills, thus raising the overall productivity of the sector. Under indirect effects, the participation of private players is supposed to increase competition in the sector, while the spread of superior technology leads to an increase in overall productivity. The thinking is that in the face of competitive pressure, inefficient public firms will attempt to raise their productivity levels, at least to be on par with those of efficient private firms, and this will increase the overall productivity of the mining industry.

Such a massive change in the investment climate of mining industry is taking place at a time of significant technological change within the industry. Not only have innovations in processing technology improved productivity and efficiency but also, by improving process control, increasing recovery rates and reducing waste, several key processing innovations have enabled firms to combine gains in competitiveness with a reduction in environmental damage (Warhurst and Bridge, 1997). Thus, it is believed that direct foreign investment through joint ventures with state-firms and/or newly privatised entities in developing countries may, under certain conditions, provide an effective vehicle for the transfer of these innovations, generating improvements in production efficiency and environmental performance (Warhurst and Bridge, 1997).

Both theoretical and empirical literature lacks consensus when examining the relationship between firm ownership and productivity. A section of theoretical and empirical economic literature argues that private firms are more productive than the public firms (Stiglitz, 1988; Majumdar, 1998; Sheshinski E and López-Calva L.F , 2003; Faria et. al., 2005; and Li and Xia, 2008). Another section of literature counters these views (Caves

[1] Raising the productivity level of the mining industry is crucial, given its distinct character. Mining is more prone to diminishing returns and increasing costs. For example, the cost of production increases substantially as the depth of mines increase. Geological characteristics also play an important role in determining productivity. Therefore, the traditional view was that productivity in the mining industry was largely determined by geological characteristics and the production cycle. However, many later studies have countered this view and emphasised the role of technology and innovation in attaining higher productivity (see Tilton and Landsberg, 1999; Aydin and Tilton, 2000). A number of productivity analyses for the mining industry have been carried out by the Centre for Study of Living Standards, Ontario, Canada. For example, see Smith (2004).

and Christensen, 1980; Boardman and Vining, 1989; and Issac et. al., 1994). The relative inefficiency of public sector firms has been explained by lack of incentive, absence of competition, and principal agent problems. The detailed theoretical explanation and empirical evidences in this respect is provided in Chapter Three which specifically deals with firm ownership and productivity issue.

Liberalisation of the mining industry also raises serious apprehension of accentuating environmental degradation and the marginalisation of local communities. The mining industry is considered to be one of the most polluting industries, especially the mining of chromites, coal and uranium. The basic textbook knowledge of public economics and environmental economics tells us that private firms only focus on the maximisation of profit, and seldom bother about the environmental health of the mining periphery. The greater participation of private firms in the mining industry therefore amplifies the apprehension of increasing environmental damage (Akabzaa and Darimani, 2001). Research by the World Wildlife Fund (WWF) shows that the relationship between the inflow of private foreign investment and the environment is complex: they can be both positive and negative. In many cases, foreign direct investment (FDI) has had largely negative impacts; especially in natural resource sectors which form the largest proportion of investment flows to the least-developed countries. WWF recognizes that the FDI can also bring substantial benefits, particularly in developing countries. However, such positive outcomes only occur inside an international regulatory framework that actively promotes sustainable development (CSD, 2000).

On environmental performance of public and private firms, economic literature is highly divided. While one set of studies point out the underperformance of private sector firms (Friedman, 1970; Baumol and Oates, 1988) another set highlights the underperformance of public sector firms, or sometimes no difference (Pargal and Wheeler, 1996). The differential environmental performance of public and private firms has been explained with the difference in regulatory compliance, internalisation of environmental damages and production efficiency. A detailed survey of the literature in this field will be provided in Chapter Four.

Mineral extraction also causes displacement of local communities, mostly the indigenous communities who are marginalised both economically and socially (Downing, 2002). There have been serious allegations on the payment of no compensation, or much less than the desired level of compensation, to the project affected households (Fernades et al., 1997). No systematic work has been done to compare the compensations provided by the public and private mining firms. Drawing arguments on the line of environmental performance of public and private mining firms we can hypothesise that public firms will provide better compensation (better social compliance) compared to private mining firms. However, better economic performance of private mining firms would enable them to provide better compensation. Therefore, the relationship between firm ownership and social compliance remains an empirical issue. A detailed theoretical framework for explaining the differential social compliance of public and private mining firms has been provided in Chapter Five.

In sum, the question whether liberalisation of the mining industry will have positive or negative impacts on the productivity, environmental health and social welfare, remains as an empirical issue. In this context, the present study empirically examines the productivity, environmental and social performances of public and private mining firms operating in India.

The Indian mining industry is an interesting case for examining the above mentioned issues for several reasons. For more than four decades, the Indian mining industry was under the unitary control of the government, with state ownership of mining firms and restrictions on private investment. With the rapid growth of the Indian economy since the early 1990s and a resultant growth in demand for mineral products, the Indian government changed its policy. With the aim of raising mineral production and the productivity of extraction, the industry was opened up to private (both domestic and foreign) sector participation in 1994. Alongside the economic liberalisation introduced by GOI in 1991, a comprehensive National Mineral Policy (NMP) was announced in March 1993. The policy introduced for the first time the idea of encouraging private investment in exploration and extraction of minerals. Recognising the lack of resources and up-to-date technology with the Geological Survey of India (GSI) for carrying out even regional (or preliminary) exploration, domestic and foreign investment was allowed into the mining industry for undertaking exploration and prospecting activities (GoI, 2006). This resulted in the entry of many private players. More interestingly, some of the Indian mining firms (both public and private) have emerged as multinational companies. Public-sector firms such as the Oil and Natural Gas Corporation (ONGC Videsh), and private-sector firms such as Tata Group, Aditya Birla Group, Essar Group, and Reliance Industries, have made forays into foreign countries for exploration, extraction and high-end value addition of minerals. At the same time, the Indian mining industry has made space for foreign private mining firms such as Insilco and Sesa Goa. Keeping in view the rapidly changing profile of the Indian mining industry, this study compares the performances of public- and private-sector firms without making any distinction between foreign and domestic firms.

1.2 Objectives of the Study

The present study seeks to examine the following research questions:

1. Are private-sector mining firms more productive than public-sector mining firms?
2. Do public-sector mining firms comply with environmental regulations better than their private counterparts?
3. Is the social compliance of public-sector mining firms better than that of private-sector mining firms?

The term social compliance has been defined as compensation made to individual households and communities as a whole for the direct and indirect economic, social and environmental losses accrued due to the involuntary land transfer.

1.3 Methodology and Data

The three questions mentioned above have been examined independently in Chapters Three, Four and Five. First, for comparing the productivity differences between public- and private-sector mining firms, total factor productivity (TFP) of the two groups have been measured for the four sectors of the mining industry—metallic, non-metallic, coal and petroleum—for the periods 1988-89 and 2005-06. For this purpose, firm-level data provided by the Centre for Monitoring the Indian Economy (CMIE) in its Prowess data base has been used. Second, for comparing the environmental compliance of public- and private-sector mining firms, the study focuses on the chromite mining industry. Data on four pollution indicators—overburden management, air pollution, the quality of mine drainage water after treatment, and the quality of ground water—provided by the Odisha State Pollution Control Board have been used for the comparison. A new methodology (adapted from poverty literature[2]) to measure environmental performance in a multidimensional framework is presented. Third, for comparing social compliance, the compensation paid by the public- and private-sector mining firms when acquiring private land is examined. For this purpose, 69 households (which make up 84 land transfer cases) which surrendered their land to public- and private-sector

[2] Alkaire and Foster(2008)

mining firms have been surveyed in the Kendujhar and Jajpur districts of Odisha. The survey, with a pre-tested questionnaire, gathered information on the direct and indirect compensations received by the households and the local community, and their perception of the land acquisition process.

1.4 Chapter Outline

The book has been organised as follows. The comparison of the performances of public and private mining firms in three different indicators—productivity, environmental performance and social compliance—need specific treatment on their own. Therefore, the review of literature, narration of the methodology and data sets used for answering the three different questions have been mentioned separately. The remainder of the book is organised on the following lines. Section 1.5 of this chapter describes the policy changes that have taken place in the Indian mining industry; Section 1.6 discusses the contribution of the mining industry to the national and state economies; Section 1.7 discusses the issues relating to resource curse in Indian context; and Section 1.8 concludes the chapter. Chapter Two provides a normative analysis of the efficiency, environmental and social implications of Indian mining laws. Chapter Three examines the first research question whether the type of ownership affects the productivity of mining firms operating in India. The differences between the environmental performances of public and private mining firms are examined in Chapter Four. Chapter Five probes the third research question of whether public and private mining firms differ in their social compliance—basically in compensating those affected by their operations. A summary of the findings is provided in Chapter Six, followed by the description of the limitations of the study and issues for further research.

1.5 Policy Changes in the Indian Mining Industry

The Mineral Policy Conference, held in 1947, led to the enactment of the Mines and Mineral (Regulation and Development) Act, 1948, which constituted the first legal framework for the regulation and development of mines in independent India. The Constitution of India empowered the Central government and the state governments to regulate mining activities and the development of minerals by including Entry 54 of List I and Entry 23 of List II in the Seventh Schedule (GoI, 2006). The Constitution of India accorded the ownership rights over mines to the states. However, to provide the benefits of randomly located minerals to all states, the Central government retained regulatory authority over all major minerals defined in Schedule A of the Industrial Policy Resolution, 1956 (IPR). They included iron ore, coal and lignite; mineral oils; the mining of iron ore, manganese ore, chrome ore, gypsum, sulphur, gold and diamonds; the mining and processing of copper, lead, zinc, tin, molybdenum and wolfram; and minerals specified in the Schedule to the Atomic Energy (Control of Production and Use) Order, 1953 (MSME, 2007). The management of the minerals listed in Schedule B, known as minor minerals, was left with the state governments. The Mines and Mineral (Regulation and Development) Act, 1957 (MMRD Act 1957) entrusted the Central government with the right to regulate mines and develop minerals. Two Rules—the Mineral Concession Rules 1960 (MCR 1960) and the Mineral Conservation and Development Rules 1958 (MCDR 1958)—were framed under the Act. While the MCR dealt with the major minerals, the state governments were free to frame their own rules for concessions to do with minor minerals. Accordingly, most states framed their own minor mineral concession rules.

The Central governments control over mining was extended in 1972 by bringing in amendments, for the first time, to the MMRD Act 1957. Measures such as premature termination of mining leases (MLs), lowering the ceiling on individual holdings, the power to modify MLs and the right of the Central government to undertake prospecting and mining operations in certain areas, removal of the ceiling on royalty charged on minerals, the inclusion of a provision to collect dead rent as part of the Act, and the enhancement of penalties were introduced. The MMRD Act 1957 was further amended in 1986, incorporating more stringent measures. For example, first schedule minerals, in which prior approval of the Central government had to be obtained under the Act, were increased in number from 27 to 38, the Central government was authorised to reserve areas for public-sector undertakings (PSUs), and mining plan approval was made compulsory (GOI, 2006).

Until 1993, the Indian mining industry thus remained under the control of the public sector, with minor private-sector participation under captive mining provisions. Private investment in the industry was minimal and in some sub-sectors, such as coal, it was almost zero. The restriction on private investment limited the scope for creating a competitive environment in the Indian mining industry.

A growing demand for minerals coupled with low productivity[3] (Jalan, 2006) made the government open up the mining industry to private (domestic and foreign) investment in 1994. Both domestic and foreign private firms were allowed to invest in the mining sector with the hope to benefit from superior technology and more capital (GoI, 2006). The National Mineral Policy 1993 brought about comprehensive changes to the country's mineral policy, including allowing private participation in the extraction of 13 major minerals which had hitherto been reserved for the public sector. The list comprised iron ore, manganese ore, chrome ore, sulphur, gold, diamonds, copper, lead, zinc, molybdenum, tungsten, nickel, and the platinum group of minerals. Consequently, the MMRD Act 1957 was amended in January 1994, and the MCR 1960 and the MCDR 1958 soon after, to incorporate the changes and simplify the procedure for granting mineral concessions. With the aim of attracting large private investments, the MMRD Act was further amended in December 1999, and the MCR and the MCDR in 2000. This brought in a number of changes in procedures for obtaining prospecting licences (PLs), reconnaissance permits (RPs) and MLs, and delegated more powers from the Central government to state governments (GOI, 2006).

Coal Industry

Apart from public-sector firms dominating coal production, government control over this sector was imposed through price controls. Administered prices prevailed in the coal sector until the late 1990s. It was partially deregulated in 1997 (grades A to D) and completely in January 2000 (grades E to G). This, in theory, conferred the right to fix the price of coal on two public-sector companies, Coal India Limited (CIL) and Singareni Collieries Company Limited (SCCL), which operate as exclusive producer-cum-traders of coal in India. However, the price fixed by the companies is, in reality, "guided" by the Ministry of Coal (MOC), Government of India (GOI, 2005).

The coal industry is dominated by two fully government-owned companies operating in two different geographical regions. These companies have never had to compete in the market place and as such have had no interest in creating a vibrant and competitive coal market

---

[3] Mining costs of Indian companies are at least 35 percent higher than those of leading coal exporting countries such as Australia, Indonesia, and South Africa. To match productivity, they will need to invest in new technologies, improve processes in planning and execution of projects, and institutionalize a comprehensive risk management framework.

(GoI, 2005, pp.57). They see their role as one of fulfilling the production targets set by the government and take up plans and projects to just meet the targets, leaving very little as a surplus to meet any unanticipated or sudden increase in demand. New players in the coal mining industry face huge entry barriers. So the supply response tends to lag, and demand-supply gaps persist. In the end, only a minuscule quantity of coal is available for free trade. To understand the demand response to domestic price variations, one has to recognise that around 80% of the domestic coal production is used for power generation (utilities plus captive) (GOI, 2005). Only 20% is left for market forces to determine the price and, of this, around 12% to 13% is consumed by brick kilns. Therefore, deregulation of the coal price even after the 1990s and 2000 has done little to ensure that it is determined by market forces.

Mineral Oil

The government has been trying to attract private (domestic and foreign) investment in the petroleum sector from as early as the 1970s, and trying out various policy instruments to realise this objective. Until the launch of the New Exploration Licensing Policy (NELP) in 1999, private-sector firms were permitted to invest only if they were part of a joint venture with public-sector firms. Public firms had the option to take a minority stake (40%) in joint ventures. After the launch of the NELP, the conditionality of joint venture was relaxed, and private firms were allowed to undertake projects independently. However, apart from this, the NELP did not provide for any of the liberal fiscal incentives that had been declared in earlier production-sharing contracts (Dey, 2001). Although the government of India proposed a production sharing mechanism in the 1970s, it did not yield much success. For example, in 1974 the government offered 7 million acres of the Bay of Bengal to Natamas Carlsburg Co. of the USA for offshore exploration and production. A contract was entered into between the US Company and Oil and Natural Gas Commission (ONGC) for the same. Subsequently another contract between Readings and Bates, USA and ONGC for Kutch basin (Gujarat) was signed. It was agreed that initially the foreign company would have a 61% share in the joint venture and the price of the crude, if produced, would be based on the Indonesian and Persian Gulf crude. 40% of the total crude would go to the US Company as "Cost Oil" towards the recovery of their expenses. 65% of the remaining crude would be ONGCs share, and the remaining 35% would go to the US Company. However, the venture was unsuccessful (Dey, 2001).

Under the NELP, private firms can undertake exploration and extraction activities independently and share a fixed percentage of the profits with the government, which is determined through competitive bidding. A part of the petrol called 'Cost Petrol' goes to the investor (say, for example, a private firm) to recover the cost, and the rest (called 'Profit Petrol') is shared among the two parties—the government and the contractor. Thus, under the NELP, public firms such as the Oil and Natural Gas Corporation Limited (ONGC) and Oil India Limited (OIL) have to compete on a level playing field with private companies for exploration blocks and new production acreage.

FDI into Mining Industry

In the first 40 years after independence, foreign direct investment (FDI) was not allowed in the mining sector. Mineral concessions were restricted to companies with less than a 40% foreign holding, as in other sectors. With the formulation of the National Mineral Policy in 1993, there was a slight easing up and FDI was allowed up to 50% with no limit on captive mines. Additional FDI could also be allowed on a case-by-case basis. All FDI proposals had to be cleared by the Foreign Investment Promotion Board (FIPB). In 1997, FDI up to 50% was taken out of the purview of the FIPB and put on an automatic approval route. For exploration and mining of diamonds and precious stones, FDI was allowed up to 74% under the automatic route in February 2000. In February 2006, the mining sector was fully opened up to FDI.

The Effect of Liberalisation

The effect of liberalisation on the industry can be observed from the steady rise in the share of the private sector in the aggregate value of minerals produced. In Table 1.1 we present the share of public sector in various sectors of the mining industry up to 2003-04, when we started our analysis. The share of the public sector in the total value of mineral production declined from 91.19% in 1988-89 to 74.61% in 2004-05. Specific industries, such as limestone, have experienced a substantial rise in private-sector participation. Similarly, the role of private firms in other sectors such as iron ore, chromite, bauxite and coal, has been growing rapidly. This is expected to reduce the government's control over the mining industry, create a competitive environment and foster productivity.

Table-1.1: Share of Public Sector in the Total Value of Mineral Production in India

Year	Baux-ite	Iron Ore	Lime Stone	Chro-mite	Manga-nese	Coal	*Mining Industry
1988-89	53.6	55.7	16.5	47.7	52.0	98.1	91.19
1989-90	51.6	55.8	15.1	40.5	48.5	98.1	90.21
1990-91	48.1	56.0	14.1	32.4	47.2	98.2	89.72
1991-92	45.8	58.4	14.1	34.7	46.9	98.3	87.72
1992-93	51.6	63.0	12.3	31.8	50.1	97.0	88.24
1993-94	51.3	59.6	12.1	31.7	58.4	97.4	89.06
1994-95	51.6	58.4	17.0	29.7	61.2	97.1	88.12
1995-96	NA	NA	NA	NA	NA	NA	88.15
1996-97	NA	NA	NA	NA	NA	NA	88.33
1997-98	50.6	55.7	9.2	38.0	56.6	98.0	85.65
1998-99	49.8	53.4	7.2	39.8	58.0	97.6	83.41
1999-00	46.0	55.2	6.3	38.7	58.8	96.0	80.43
2000-01	42.0	53.9	7.7	32.1	59.5	95.8	80.68
2001-02	46.5	52.3	7.3	20.8	55.4	95.3	79.45
2002-03	52.9	50.2	6.5	23.0	55.5	95.4	76.86
2003-04	49.9	46.8	6.4	26.7	53.4	94.7	74.96
2004-05							74.61

Source: Indian Mineral Year Book, Various Volumes; * Source: Indian Mineral Industry At a Glance Various Volumes, Government of India, Ministry of Mines, Indian Bureau of Mines, Nagpur.
Note: NA- Not available

1.6 Contribution of the Mining Sector to the National and State Economies

Before entering into the firm wise analysis it is pertinent to have a broad understanding of the mining sector in India. For this purpose we use the data on the mining and quarrying sector (M&Q) published in the website of the Reserve Bank of India in its section Handbook of Statistics on Indian Economy. Table-1.2 presents the value of output in the M&Q sector from 1988-89 to 2005-06. Also, we compute the share of the M&Q sector in the gross domestic product (GDP) of the country, annual growth rate and average growth rates for different periods of the mining industry. At the national level, the M&Q sector has a very small share in the GDP; it has

Table-1.2 Growth of Mining and Quarrying (M&Q) Sector in India

Year	Value of output*	Share of M&Q sector in GDP	Annual growth rate %
1988-89	39433	2.5	16.17
1989-90	42429	2.6	7.60
1990-91	46868	2.7	10.46
1991-92	48442	2.7	3.36
1992-93	48888	2.6	0.92
Average Growth rates from 1988-89 to 1992-93			7.70
1993-94	49568	2.5	1.39
1994-95	54171	2.6	9.29
1995-96	57349	2.5	5.87
1996-97	57667	2.4	0.55
1997-98	63324	2.5	9.81
1998-99	65114	2.4	2.83
1999-00	67190	2.3	3.19
Average Growth rates from 1993-94 to 1999-2000			4.70
2000-01	68797	2.3	2.39
2001-02	70002	2.2	1.75
2002-03	76194	2.3	8.85
2003-04	78549	2.2	3.09
2004-05	84954	2.2	8.15
2005-06	86083	2.6	1.33
Average Growth rates from 2000-01 to 2005-06			4.26
Average Growth rates from 1988-89 to 2005-06			5.39

Source: RBI (2011) Handbook of Statistics viewed on website on 11th May 2011
Note: * At factor cost in rupees crore at 2004-05 base year price

remained around two percentages. The annual growth rate of M&Q sector shows a high rate of fluctuation. The average growth rate of the sector in the pre-liberalisation period (1988-89 to 1992-93) was highest at 7.70%. Gradually the growth rate has slowed down in the first (1993-94 to 1999-2000) and second phases (2000-01 to 2005-06) of liberalisation to 4.7% and 4.26% respectively.

Table 1.3 presents the share of the M&Q sector as a percentage of the net state domestic product (NSDP) in 21 states from 1993-94 to 2007-08. Here, our analysis is confined to the major mineral-producing states only. In a majority of the states, the M&Q sector has a very negligible share in the NSDP. Only in a few states such as Chhattisgarh, Jharkhand, Meghalaya, Odisha, Assam, Goa, Andhra Pradesh and Madhya Pradesh does the M&Q sector have a sizeable share in the NSDP. In 2007-08, Chhattisgarh recorded the highest contribution of the M&Q sector to its NSDP at 14.43%. All other states recorded less than 10%. Jharkhand ranked second in the contribution of the M&Q sector to its NSDP at 8.66%. This was followed by Meghalaya (8.35%), Odisha (6.39%), Goa (4.13%), Madhya Pradesh (3.90 %), Andhra Pradesh (3.56%) and Assam (3.29%). Inter-temporal analysis of the share of the M&Q sector in the NSDP reveals that a few states have recorded steady growth while others have experienced a decline. In the former category are Andhra Pradesh, Chhattisgarh, Madhya Pradesh, Meghalaya, Odisha and Rajasthan. In the latter are Assam and Jharkhand. Goa has shown a fluctuating trend; between 1993-94 and 2000-01 the share of the M&Q sector in its NSDP fell from 5.58% to 3.16% but then recovered to rise to 4.13% in 2007-08.

The contribution of the mining sector to the economy can also be assessed by looking at the revenue generated from mining activities. Mineral-rich states derive a sizeable amount from this source. A major component in this is the royalty collected from mining firms. In addition, states collect dead rent from lessees who have not been operating their mines and thus not paying any royalty. Besides royalty and dead rent, states also get an income from the initial application fee payable by a concession seeker, the annual fee payable by RP/PL holders on the basis of the area held, surface rent, sales tax or value added tax (VAT), local area tax (for example panchayat tax), and stamp duty. Some states, for instance Odisha and West Bengal, have also imposed a specific cess and surcharges on minerals to mobilise additional revenue for special purposes. However, revenues from all these sources are meagre, even in comparison with the modest returns from royalty and dead rent. No systematic data is available on the total revenues collected by states from mining sources. Besides, royalty being the chief source of revenue from mining, we calculate its contribution to the total revenue receipts (TRR) of 16 major states. Table 1.4 presents the total amount of royalty collected by the states and its share as a percentage of the TRR between 2002-03 and 2004-05. Jharkhand received the largest amount as royalty from mining activities, followed by Andhra Pradesh, Madhya Pradesh, Chhattisgarh and Odisha. In terms of the percentage of contribution to the TRR,

Jharkhand led with 12.54%, followed by Chhattisgarh, Odisha and Rajasthan with shares of 9.31%, 5.77% and 3.37 % respectively.

Table 1.3 Percentage Share of Mining and Quarrying Sector in NSDP

States	1993-94	1995-96	2000-01	2005-06	2007-08
Andhra Pradesh	2.06	1.72	2.51	3.04	3.56
Arunachal Pradesh	1.42	0.38	1.42	1.16	0.97
Assam	5.97	4.29	4.56	3.58	3.29
Bihar	0.12	0.09	0.21	0.11	0.08
Chhattisgarh	7.80	7.17	13.36	14.55	14.43
Goa	5.58	4.72	3.16	3.77	4.13
Gujarat	2.63	1.95	2.45	1.93	NA
Haryana	0.20	0.15	0.28	0.35	0.28
Jharkhand	15.50	13.48	12.81	10.30	8.66
Karnataka	0.61	0.54	0.57	0.90	0.95
Kerala	0.21	0.18	0.29	0.49	0.59
Madhya Pradesh	2.78	2.32	3.11	3.46	3.90
Maharashtra	0.52	0.50	0.79	0.75	0.63
Meghalaya	3.38	3.50	7.92	8.04	8.35
Odisha	4.33	4.25	4.60	6.14	6.39
Punjab	0.01	0.01	0.00	0.02	0.02
Rajasthan	1.93	1.84	2.06	2.27	2.69
Tamil Nadu	0.59	0.46	0.41	0.47	0.43
Uttar Pradesh	0.63	0.73	0.87	1.08	0.79
Uttarakhand	1.40	0.96	0.57	1.34	NA
West Bengal	1.20	1.02	1.16	1.18	0.93

Source: Computed from the data on the Reserve Bank of India (RBI) website (viewed on 9 July 2008 and 2 November 2009) on components of NSDP at factor cost

Table-1.4 Share of Royalty in the Total Revenue Receipts of Major Mineral Producing States

(Royalty in Rupees crores)

States	2002-03		2003-04		2004-05	
	Royalty	R as a % of TRR	Royalty	R as a % of TRR	Royalty	R as a % of TRR
Andhra Pradesh	769.93	3.35	766.56	2.85	864.53	2.70
Assam	9.36	0.14	12.64	0.16	13.36	0.10
Chhattisgarh	552.36	10.20	637.17	10.69	694.61	9.31
Goa	14.81	0.81	17.87	1.10	17.44	0.90
Gujarat	172.63	0.97	217.90	1.19	238.95	1.18
Haryana	118.08	1.36	76.77	0.78	92.50	0.81
Jharkhand	797.65	10.77	900.16	12.09	916.20	12.54
Karnataka	83.89	0.52	143.62	0.69	210.94	0.83
Kerala	1.63	0.02	10.45	0.09	12.61	0.09
Madhya Pradesh	590.69	4.41	646.71	4.53	733.72	3.58
Maharashtra	400.69	1.29	475.92	1.38	568.24	1.33
Odisha	440.57	5.22	547.20	5.80	663.61	5.77
Rajasthan	399.68	3.06	457.96	2.97	589.79	3.37
Tamil Nadu	297.34	1.43	324.50	1.37	324.82	1.20
Uttar Pradesh	262.42	0.94	254.18	0.80	291.94	0.76
Uttaranchal	22.55	0.70	30.65	0.85	35.60	0.72

Source : Budget documents of the respective State Governments
https://59.160.162.25/businessobjects/enterprise115/desktoplaunch1/InfoView/main/main.do?objId=6169
Retrieved from RBI website on July 09 2008
Note: Royalty; TRR: Total Revenue Receipts

From the above two indicators (share of royalty in the TRR and share of M&Q in the NSDP), it is clear that the states where mining is a major economic activity are Jharkhand, Chhattisgarh, Odisha and Madhya Pradesh. In spite of possessing a huge mineral base, these states have remained the most backward in India, with low economic growth, per capita income and human development.

1.7 Resource Curse in the Indian Context

It has been observed across the world that richly natural resource-endowed countries have remained backward, both in economic growth and human development. The backwardness of the resource rich regions has been conceptualised as 'resource curse' in economic literature (Auty, 1993; Sachs and Warner, 1995; Kalshian, 2007; Akanni, 2007; Andersen and Aslaksen, 2008; Brunnschweiler, 2008; Lederman and Maloney, 2008; Cavalcanti, Mohaddes and Raissi, 2009). This literature points out that a favourable natural resource endowment may be less beneficial to countries at low- and mid-income levels of development. Auty, (1993, pp.1) defines resource curse as '*not only many resource-rich countries fail to benefit from a favourable endowment; they may also perform worse than less -endowed countries*'. In some places it breeds a high level of corruption, and in others it leads to currency appreciation, which in turn adversely affects domestic exports. Apart from depressing macroeconomic performance, mining operations inflict deep wounds on local economies. Not only do the local communities in mining areas remain poor, their vulnerability increases due to displacement and environmental degradation (CSE, 2008).

In India, mineral-rich states such as Jharkhand, Chhattisgarh and Odisha have low per capita incomes compared to states that are less endowed with mineral resources. They also have higher levels of poverty, lower growth rates and higher levels of infant and maternal mortality, malnutrition and morbidity. A study carried out by Srivastav (2006) under the auspices of the World Bank attributed the resource curse in Odisha to inappropriate institutional arrangements to devolve resources to local governments for the development of communities in mining peripheries. The resource curse is thus very much a reality in the mineral-rich areas of the country. Additional evidence comes from the fact that 60% of the 50 major mining districts figure among the 150 most backward districts in the country. Thirteen of them are among the 50 most backward districts and four of them—two in Odisha and one each in Jharkhand and Chhattisgarh—are among the 25 most backward districts in India (CSE, 2008; p. 15).

The resource curse is evident even in a micro-level analysis. The World Bank study mentioned earlier focussed on two blocks in mineral-rich Kendujhar district of Odisha—Joda, which has a high concentration of mines, and Kendujhar Sadar, which then was likely to be mined intensively in the near future. The study revealed that the households in Kendujhar Sadar were significantly better off in terms of average cash income and ownership of productive assets compared with those in Joda. Education levels were also higher in Kendujhar Sadar households, while those in Joda reported higher incidences of family illness. Wage income was higher in the households in Joda, most probably because of employment opportunities in mines in the locality. However the difference was not statistically significant. The study concluded that the proximity to mines was detrimental in a number of ways, including greater environmental damage and health hazards (Srivastav, 2006).

1.7.1 Mining and Forest Degradation

Another major paradox of mineral endowment is that a large part of this wealth is in areas that have dense forest coverage. India's major mineral-producing districts are characterised by thick forests, a large adivasi population and a high incidence of poverty and backwardness. Most of the potential mines are in currently forested areas, which may not be evident by looking at forests in mineral producing districts where forests are already depleted. The average forest cover of the 50 major mineral-producing districts is 28%; forests cover 11,890, 400 hectares (ha) in them. This is 18% of the total forest cover in the country. Out of these 50 districts, in six districts more than 50% of the geographical area is covered under forest. About 62% of the 50 districts have forest cover that is more than the national average of 20.6% (CSE, 2008). The forest cover of the top five mining states—Andhra Pradesh, Odisha, Chhattisgarh, Jharkhand and Madhya Pradesh—is above the national average. Chhattisgarh has the highest forest cover of around 43%. Jharkhand has forests on 30% per cent of its area, while Odisha and Madhya Pradesh have forests on 27% and 26% of their areas respectively.

Due to the geographical specificity of mineral endowments and the priority the government gives to mineral extraction, large patches of forests are cleared every year. In a press release issued on 7 January 2010, the MoEF declared that since the enactment of the Forest Conservation Act (FCA) in 1980 a total of 1309 cases have been approved for forest diversion under various categories of mining, which amounts to 100,871 hectares of forests lost (India Together, 2010). This figure would be even higher if we took into account the forest land diverted before 1980 when many coal mines took over vast areas of land—mostly forests (IT, 2010).

To control the depletion of forests, the Central government has made it mandatory for all mining firms to submit an Environmental Management Plan (EMP), in accordance with which they are bound to undertake compensatory afforestation. Nevertheless, the compensatory afforestation programme continues to be a fiasco. In many cases, state agencies do not take any interest in carrying out compensatory afforestation programmes.

The audit report on Madhya Pradesh by the Controller and Accountant General (CAG) for the year ending on 31 March 2007 indicated that the law had been violated by the state's forest department, which granted permission for massive diversion of forest land. Since the Forest (Conservation) Act came into force in 1980, Madhya Pradesh has diverted 51,018 ha of forest land for non-forest purposes, or 734 projects. As per the provisions of the Act, the state needed to carry out compensatory afforestation on 73,213 ha of land but an audit scrutiny of the records in the nodal office revealed that, as of June 2006, no afforestation had been carried out in the case of 289 projects (a 39% shortfall) and on 13,441 ha of stipulated land (an 18% shortfall) while Rs. 82.60 crore, recovered from user agencies, had not been utilised (a 75% shortfall in utilisation of funds) (IT, 2008). Similarly, in Dhenkanal district of Odisha, 1,054.843 ha of land were diverted for various development projects, including mining. Although the state forest department, during 2003 and 2008, deposited a total of Rs. 76.4 crore with an independent agency named the Compensatory Afforestation Fund Management and Planning Authority (CAMPA), not a single tree had been planted in the district until April 2008 (Dharitri, 2008).

Box 1.1: List of sanctuaries where mining has been allowed

1. Jamwa Ramgarh Wildlife Sanctuary in Rajasthan has been extensively mined for marble and soapstone.
2. Keladevi Wildlife Sanctuary, which is part of the Ranthambore Tiger Reserve in Rajasthan, has been affected by red sandstone and limestone mining.
3. Bhagwan Mahaveer Wildlife Sanctuary in Goa has at least 100-150 mines within a 10-km radius of the protected area. Mine leases are spread over more than 40% of the forest area in Goa.
4. Gangua Sanctuary in Madhya Pradesh is extensively mined for diamond and white sandstone.
5. Areas of Darlaghat Wildlife Sanctuary in Himachal Pradesh were de-notified to allow limestone mining.
6. Gir Wildlife Sanctuary and National Park in Gujarat, the last home of the Asiatic lion, has 100-odd mines within a 10-km radius of the protected area.
7. The Gujarat government de-notified Narayan Sarovar Wildlife Sanctuary reducing its size from 766 sq km to 444 sq km to allow limestone and lignite mining.

Source: CSE, 2008; p. 75

While permitting mining projects, even environmentally sensitive areas have not been spared. Throughout the country, mining is taking place inside reserve forests, protected forests, national parks and wildlife sanctuaries. Box 1.1 provides the list of the wildlife sanctuaries where mining operation have been allowed. The rate of deforestation due to mining activities has also increased over the years. Between 1980 and 1997, clearance was granted to 317 mines, diverting 34,527 ha of forest land. However, between 1998 and 2005, the MoEF cleared 881 mining projects in forest areas, diverting 60,476 ha of land. Thus, the total forest area diverted every year for mining purposes, between 1998 and 2005, was four times higher than that during the years from between 1980 and 1997 (CSE, 2008; p. 74).

Apart from forest degradation, mining activities leave their footprints in a number of other ways, such as impact on land, water, air, wildlife and ecology, noise pollution and health hazards. Mining activities around the world have been accompanied by land expropriation and environmental degradation that harm the livelihoods and health of local communities (Keenan et al 2002; Sosa 2000). Mining can acidify the soil and water, increase toxic chemical availability, and increase siltation of water and leaf surfaces. These effects in turn are known to decrease water availability, decrease plant growth, and as a result, decreased wildlife abundance and diversity (Saxena et al. 2002; Rasmussen and Koroleva 2003). Irrespective of the type of mining—open cast or underground—mining invariably results in enormous land disturbance—e.g. large scale excavation, removing of top soil, dumping of solid wastes, cutting of roads, creation of derelict land, etc. Surface or open cast mining has more potential impact on land than underground mining. Although underground mining has considerably less environmental and social impact than open cast mining on land, it causes enough damage through subsidence as observed in the Jharia and Ranigunj coal fields (Krishnamurthy, 2004). The surface subsidence inflicts severe damage to engineering structures such as highways, building, drainage and bridges, besides interfering with the ground water regime. Mining requires large quantities of water for dust control, fire protection and coal washing. When this water drains through a large area of the mine, it carries with it many soluble minerals that may be present either in the coal or associated rocks, thus causing severe degradation of water quality

Air pollutants originate from various activities of mining such as drilling, crushing, screening, blasting, coal washing, dust from roads, mine-waste piles and stock piles, exhaust fumes from equipments, vehicular traffic and truck haulage. Transportation of coal from mines to plants is another major source of dust. The cumulative effects of these mining activities produce enormous noise and vibrations in the mining area which constitutes a source of disturbance. Unloading, transportation and storage of minerals also cause enormous dust pollution in the mining areas.

Freshly extracted bauxite ore remains chemically very active and is easily attacked by atmospheric oxygen, humidity, heat, light and micro-organisms. Rains dissolve metals from the ore pile and the acidic solution infiltrates into the ground. Similarly, water run-off from mines including the one used for cooling drill bits and dust control and possibly carrying toxic substances used in ore processing, infiltrates into the soil. A part of it may reappear on the surface joining a water channel (river, nala etc); another part may join the groundwater reservoir. Thus both the ground water and surface water sources stand in real danger of pollution, and metalicious and acidic solutions can sometimes result in disruption of well-established eco-systems, with wide ranging repercussions (Lal, V. B. et al., 1988).

Deeper excavation on the surface or underground digging causes the water table to sink locally, often drastically, resulting in the wells and springs of the neighbourhood area going dry. In the hilly terrain, the process of unearthing and attendant landslides frequently expose the passage of underground water, thus depriving the spring of the water (Valdiya, K. S., 1988).

1.7.2 Mining, Adivasis and Displacement

Mining industries across the world have caused a great deal of displacement among local communities (World Bank, 1993). Most of these displacements are involuntary, and underpayment of compensation for their economic and social losses has been widespread (Cernea, 1999; 2003). Keeping in view the profit maximisation objective of private firms, it is alleged that the greater participation of private players will further the marginalisation of local communities. The opposition to private sector mining is witnessed through the growing naxalism in the mining areas.

A large section of India's adivasi population—numbering more than 8.4 crore, which is approximately more than 8% of the total population—inhabits in mining areas. About 90% of India's coal and 80% of its other minerals are found in adivasi areas (Jha, 2006). Of the 50 major mining districts of the country, almost half are adivasi districts (CSE, 2008; p. 9). As most adivasis inhabit in the forest areas, their livelihood and economy are closely intertwined with the fate of forests and water resources. Forest degradation due to mining and other development projects has significantly depleted the eco-system, rendering the adivasi population more socially and economically vulnerable.

No reliable data is available on the total number of people displaced due to development projects. Fernades et. al. (1997) provide an account of displacements between 1950 and 1991, which shows that mining has displaced the second highest number of people after dams. Between 1950 and 1991, mining projects displaced around 25.5 lakh people. See Table 1.5 for a detailed account of displacements and their rehabilitation. More importantly, not even 25% of the displaced people have been resettled. These figures are only for people who were moved from their land; they do not include the thousands who were dependent on the land for their livelihoods, or those whose lives were affected by the disruption of water tables, dumping of the overburden on fertile agricultural land and the destruction of forests. The worst victims of mining projects have been the adivasi population. Of the total number of people who have been displaced by various development projects, about 41% are adivasis. In the case of mining projects, about 52% of the people displaced were adivasis (Fernades et. al, 1997).

No systematic data is available on the number of people displaced due to mining after 1991. Nevertheless, one can guess the volume of displacement by looking at data on the number of mining projects that were approved after 1991 and the area of forest land diverted for mining. Between 1995 and 2005, about 74,000 ha of forest land were diverted for mining projects. Further, mining-induced displacement must have gone up substantially since the 1970s because India's coal production shifted from underground mines to predominantly open-cast ones. Many studies have reported on the inadequate rehabilitation measures for displaced households and underpayment of compensation to them (Bhengara, 1996; Panda, 2007).

1.8 Conclusion

The liberalisation of mining industries across the world has been carried out with a view to raising total production and productivity levels by attracting private (both domestic and foreign) capital. Liberalisation is expected to raise the overall productivity through a higher level of competition and the presence of more efficient firms. The underlying assumption is that private mining firms are more productive than public ones. Nevertheless, this raises the apprehension that greater participation of the private sector in mineral extraction may accentuate environmental degradation and further marginalise local communities. This study seeks to verify the veracity of the above concerns by undertaking an empirical analysis of the mining industry in India. It attempts to compare the productivity, environmental and social performance of public- and private-sector mining firms operating in the country.

Table 1.5 Mining-Induced Displacement during 1951-1990

Types of projects	All DPs (lakh)	% of total DPs	DPs resettled (lakh)	% of resettled DPs	Adivasis displaced (lakh)	% of all DPs	Adivasis resettled (lakh)	% of adivasis resettled
Mines	25.5	12	6.30	24.7	13.30	52.20	3.30	24.81
All devt. projects	213	100	53.90	25	95.39	40.9	21.16	24.78

Note: DP- Displaced persons; devt.-development
Source: Fernades et. al. (1997)

CHAPTER TWO

EQUITY, EFFICIENCY AND ENVIRONMENTAL IMPLICATIONS OF INDIAN MINING LAWS

2.1. Introduction

In this chapter we shall analyse, theoretically, the equity, efficiency and environmental implications of Indian mining laws. The basic objective here is to theoretically examine the pros and cons of the disparate alternatives to mineral ownership, the ownership of extractive mining firms and their implications for productive efficiency, environmental impact and effect on local communities. Section 2.2 of this chapter looks into the efficiency implications of various forms of mineral ownership and ownership of extractive firms. In section 2.3, we examine the issues to do with the simultaneity of mineral rights and surface rights, the determination of an optimal compensation scheme, and how the ownership structure influences the level of compensation. Section 2.4 briefly describes the existing regulatory mechanism for checking environmental damage in mining areas and its ineffectiveness. The question of whether the type of firm ownership has an influence on environmental performance is then analysed. Section 2.5 provides a summary and conclusions.

When undertaking mining projects, environmental protection and local community development deserve as much priority as profit maximisation. An overemphasis on profits, overlooking the two other dimensions, could make a project socially unviable. Damage to the local community and environment breeds social opposition, which could threaten the basic economic viability of a project. The mining laws of a country should therefore ensure that (i) mining firms have a competitive and conducive environment to maximise output (revenue) per unit of input (cost); (ii) local communities attain a better standard of living, or at least that their living conditions are not worsened; and (iii) the ecological balance is safeguarded.

2.2. Efficiency Implications

Given that mineral resources are non-renewable, scarcely endowed and occur at specific locations, a primary concern is that their ownership confer the maximum benefit on the welfare of a nation. Alongside this is the concern that their extraction be carried out in the most efficient manner. Do different ownership structures over minerals and extractive firms have different efficiency implications? In this section, we consider each of these concerns and weigh their merits. In the first part, we discuss the welfare and strategic implications of mineral ownership, and in the second part, we deal with the implications of firm ownership on extraction, or productivity.

2.2.1 Efficiency Implications of Mineral Ownership

Who should own mineral resources? Individuals, households, communities, or any tier of government such as panchayats, districts, states or the Central government? Which ownership structure over minerals will be socially efficient? Answers to these questions will not have universal application. A different set of political, legal and socio-economic factors will determine the efficient ownership structure for different nations. Here, our discussion on the efficiency implications of mineral ownership will be confined to Indian context.

2.2.1.1 Open access?

Can we allow open access to mineral resources? Open access to minerals would result in their rapid depletion due to an extraction rate much higher than the socially optimal rate (Gordon, 1954; Hardin, 1964; Baland and Platteau, 1996[1]). Unfettered extraction of minerals would also accentuate environmental problems. Given that a large part of the mineral stock is beneath public land, to which access is de facto open, the scope for overexploitation is very high. In many developing nations, including India, there are already instances of illegal mining operations on public land (Lahiri-Dutta, 2007). Open access to minerals would, therefore, be socially and economically unwise.

[1] These studies examine the over-extraction of renewable resources in open-access situations. Their findings are undoubtedly relevant to all open-access situations. The problem might be different with non-renewable resources because their extraction cost is high. Nonetheless, easily accessible resources will be depleted fast.

2.2.1.2 Individual/Household ownership?

Can we extend mineral ownership rights to surface right owners? Doing so may lead to a competitive market for mineral rights but will have a negative fallout on mineral extraction. The location of minerals is uncertain. It is likely that a place where a substantial stock of mineral resource is discovered may have a large number of small, private landowners. All individual landowners might not be interested in extraction and further value addition. For, it would be uneconomical for owners of small pieces of land to extract a mineral, and they would therefore prefer to accord priority to surface use. It could also be that none of the landowners show any interest in mineral extraction but other individuals or firms do. In such cases, tying up mineral ownership with landownership or surface rights may be inefficient.

In any case, allowing each landowner to operate on a small scale would not be an efficient solution. There are economies of scale in mineral extraction. Thus, many small operators trying to extract minerals beneath their small pieces of land may not turn out to be efficient. It is more efficient to extract minerals on a large scale from a sufficiently large area of land. Efficient extraction of minerals needs a definite scale of operation. Marginal land holdings will be obstacle to attain the minimum scale of operation. Although, cooperative mineral extraction could be a possibility, this could encounter the usual problems of common action. A large group and group heterogeneity either leads to failure of common action or results in a sub-optimal payoff (Olson, 1965; Ostrom, 1990). In the case of small land holdings, the necessary scale of operation can be attained by the transfer of ownership or user rights. However, market mechanisms may not be successful in enabling the transfer of user or ownership rights. Holding out could prove to be the main obstacle to attaining a minimum scale of production. Miller (1973) cites reasons why landowners may not desire to lease out their land. For a landowner, the expected value of leasing will be higher than the rent that a mining firm would be interested in paying. As long as the landowner is confident that exploration will take place, the land is worth more to him than the average amount a mining firm can afford to pay.

Apart from scale of operation, negative externalities can be formidable obstacles to mineral extraction. Extraction on one piece of land may have serious negative effects on the neighbourhood; for example, depletion of ground water and its pollution, air pollution, noise pollution, damage to the buildings in the case of blasting, and so on. Unless these externalities are adequately compensated for they may generate social disputes and escalate the transaction cost of the project.

All these factors make a strong case against individual/household ownership of minerals or conferring mineral ownership rights on surface right owners.

2.2.1.3 Which tier of government?

Our next alternative would be the government ownership of minerals. Then the question is which tier of government should retain ownership rights? To answer this question, we need to think about ease of administration and attaining economies of scale. If local-level governments are given mineral ownership rights in a federal set up like India's, it may aggravate competition among the states to provide tax exemptions and other incentives; the ultimate losers being the state exchequers. Though this will have a favourable influence on the private efficiency, it will have serious negative implications for equity. At the same time, Central ownership would complicate the administrative process. Moreover, the Indian Constitution vests ownership rights over land in the state governments.

In a federal set up where sub-national entities are randomly endowed with different natural resources, access to resources and their management remain controversial issues. It is obvious that multiple ownership—private or community—of these resources would significantly increase the transaction cost. In extreme cases, it could also hinder the smooth flow of resources within the country as well as between certain key sectors. For example, thermal power generation in India accounts for more than 70% of the total power output and coal remains the primary fuel of these power plants. If the property right over coal passes into the hands of private individuals or local communities, it might cause a serious disruption to the supply of coal and hence power generation. State ownership of resources reduces the uncertainties associated with other property-right institutions and pays heed to the strategic concerns of the nation. Thus, the rationale for sole (government) control over minerals holds that it minimises the transaction cost.

Another important argument for the government ownership of minerals is the need to achieve sustainable development. Both private and community ownership of mineral resources would be incapable of ensuring this. The insensitivity of market forces to distributive justice, both intra-generational as well as inter-generational, is well recorded in the literature on public economics. Community ownership of non-renewable resources would also fail to take into account the needs of future generations. Under these circumstances, state ownership of mineral resources appears to be an ideal solution.

Government ownership of mineral rights allows for direct accountability of the social price for mineral-bearing land and the external costs (Long, 1995). Further, national oversight ensures that the relative global availability of minerals and other values are considered. It can also promote adaptive efficiency by publicising creative local solutions, providing technical support, and funding useful research. Nevertheless, Long suggests that local evaluation and planning are necessary to mitigate environmental costs and resolving land-use conflicts.

2.2.1.4 Market Structure, Output and Price

Another question that arises has to do with the nature of the market and its efficiency implications. Should the mining industry be competitive or monopolistic? What would be the implications for price and output under different market structures?

In competitive equilibrium, an individual is indifferent to selling a unit of extractible resource either today or tomorrow. A monopolist, on the other hand, compares the marginal revenue he obtains in this period with the discounted marginal revenue he would obtain in the next period by transferring a unit of sales from this period to the next. If we have constant elasticity demand schedules, then the price will be proportional to marginal revenue, and the two equilibria will be the same[2]. Stiglitz (1976) concludes that this would give only a very limited scope for the monopolist to exercise his power. In other cases, there is a tendency of the monopolist being more "conservation minded" than the competitive market (Solow, 1974). Solow does not endorse any centralised decision making, a process that is likely to have imperfections and externalities of its own. Indeed it might be enough to have the government engaged in a continuous programme of information-gathering and dissemination, covering trends in technology, reserves and demand.

Hotelling (1931) argues for the government ownership of minerals to control the windfall gains from new discoveries of minerals. He holds the

---

[2] If the elasticity of demand in the next period is higher than in this period, the ratio of marginal revenue to price will be higher in the next period than in this period, which means that at a competitive price, discounted marginal revenue in the next period exceeds marginal revenue in this period. So it pays to sell more in the next period. The monopolist is more conservationist than the competitive market. The process is reversed if the elasticity in the next period is lower than in this period.

view that the unexpectedness of mineral discoveries provides a good enough reason for governmental control and special taxation. Great profits of a thoroughly adventitious character arise in connection with mineral discoveries and it is not good public policy to allow such profits to remain in private hands (p. 144).

Governments owning mineral rights have a conflict of interest between their roles as a profit-maximising landowner and as a guardian of public welfare. As a monopoly supplier, governments have considerable power to manipulate mineral-rights markets. To avoid monopoly rent-seeking by governments, a competitive market for government-owned mineral rights must be created by artifice (Long, 1995; p. 74).

A competitive market with many owners of mineral rights and many mining firms should be adequately served by laws that guarantee enforceability of contracts, compensation for damages, and regulations to mitigate external costs. In most countries, however, the government is the major or sole owner of mineral rights. Although this makes the government a monopoly supplier of mineral rights, a competitive market for mineral rights can be created by artifice. By conveying mineral rights on demand to a large number of mining firms, these rights are effectively parcelled out among many "owners". Having de facto private property rights, firms are then able to trade these mineral rights among themselves, to secure financing, to enforce contracts, and to obtain compensation if mineral rights are later condemned for public use (Long, 1995; p 76).

We have made a strong case for the government ownership of minerals and also dropped the alternative of giving the right to local governments. The remaining options are therefore either state or Central government ownership. It would be difficult for the Central government to efficiently monitor mineral resources. Central government ownership may not be efficient for administrative purposes as well. Though local governments would be most efficient for monitoring purposes, it would give rise to other problems, as mentioned in an earlier section. For paying attention to strategic concerns, ensuring the free flow of minerals, and implementing a uniform policy throughout the country, Central regulations would be most efficient. However, it would be socially efficient to give state governments the ownership right over minerals. Except for a few important regulatory roles such as implementing a uniform mineral policy, formulating a trade policy for minerals and granting environmental clearance, the Central government should not intervene much in the management of the minerals. The author has elsewhere (Das, 2009) argued that state governments should be provided with more elbow room to determine their own royalty rates.

2.2.2 Firm ownership and extraction efficiency

After analysing the efficiency implications of the nature of mineral ownership it is pertinent to scrutinise the efficiency implications of firm ownership on extraction efficiency. Even if there is state ownership of mineral resources, who—state-owned enterprises (SoEs) or private-sector companies—can extract minerals in the most productive manner?

That the government owns mineral resources does not mean that only SoEs can extract them efficiently. The government can allow private companies to extract resources under licensing or competitive bidding. For social efficiency, it is immaterial whether the government demands a royalty or not (though it may have an impact on public finances), and having more public finances may be useful if the opportunity cost of public resources is high. How should mining companies obtain surface rights from landowners? Can public and private companies behave differently in this regard? Why should they behave differently? Can the institutional environment (for example, the weak bargaining power of landowners) be a reason for such different behaviour? Who—SoEs or private companies—should extract mineral resources? Given that we have argued for the states to be granted the ownership of minerals, two possibilities are left; either SoEs extract the minerals, or the user rights are leased out to private companies that will carry out extraction. Under the second alternative, issues may arise in relation to contracts between the owner of the mineral, or the surface right owner and the private company that has secured the lease, creating uncertainties. Such uncertainties may have productivity implications. Therefore, we consider the following issues.

2.2.2.1 Extraction contracts and productivity

Many of large projects need relationship-specific investments (RSI). For example, a mineral extractor may have to make investments in each piece of land where extraction is to take place. This may make the contract between the surface right owner and the mineral extractor complex. What if one of the parties wants to renegotiate the terms of the contract after the other party has made RSI? What if each party takes "excessive precaution" against such a contingency? Does this, in some situations, discourage the parties from entering into a contract or deal? Or does this reduce the surplus that can be potentially generated from such collaboration? Are there obstructions to contracts between government and private companies on the use of mineral resources? Does the need for RSI encourage one of the parties (for example, the public) to demand renegotiation of the terms of contracts in its favour? (The potential of the state taking over assets created by multinational companies for petroleum extraction in some countries, as in Latin America, can be an indication.) Does this prevent private companies from entering into contracts? Though private companies may take precautions in this regard, empirical evidence from India and many other countries does not show that they are averse to extracting minerals after obtaining licences from the government. Thus there seems to be no need of extraction by the public sector on the grounds that a regime of private extraction under licensing will yield sub-optimal results.

According to transactions cost theory, when exchange involves significant investments in relationship-specific capital, an exchange relationship that relies on repeated bargaining is unattractive (Klein et. al., 1978; Williamson, 1979; 1983; 1985). Once investments are sunk in anticipation of performance, "hold-up" or "opportunism" incentives are created ex post which, if mechanisms cannot be designed to mitigate the parties' ability to act on these incentives, could make a socially cost-minimising transaction privately unattractive at the contract execution stage (Klein et. al., 1978). A long-term contract that specifies the terms and conditions for some set of future transactions ex ante provides a vehicle for guarding against ex-post performance problems (Joskow, 1985; 1987). Joskow (1987) provides both theoretical reasoning and empirical evidence (from a study of 277 coal contracts in the US) to show that as RSI become more important, the parties will find it advantageous to rely on longer-term contracts that specify the terms and conditions of repeated transactions ex ante, rather than relying on repeated bargaining.

Tang (2009) states that there is no doubt that a long-term contract saves renegotiation costs and induces more efficient specific investments, but it also reduces the flexibility of the contracting parties when facing future contingencies. It is therefore important to apply appropriate pricing provisions to maximise the expected revenue generated from the long-term relationship, subject to the flexibility constraint. For example, if a fixed price is negotiated for trading a certain good over multiple years, it will not be easy for the contracting parties to accommodate future contingencies. However, if a price-adjust mechanism is included in the contract, such as designing the price as a function of the current consumer price index, the contract price will automatically reflect the current market situation. Tang's empirical results show that coal contract durations increase with RSI and decrease with uncertainty.

Klein, Crawford and Alchian (1978) maintain that the crucial assumption underlying [the RSI analysis] is that as assets become more specific and more appropriable, quasi-rents are created (and therefore the possible gains from opportunistic behaviour increase). So the costs of contracting will generally increase more than the costs of vertical integration. Hence, ceteris paribus, we are more likely to observe vertical integration

Miwa and Ramseyer (2000) visualise a scenario where production requires large idiosyncratic investments, and detailed contracts are not feasible. In such a world, production would generate quasi-rents, and the quasi-rents would in turn create the risk of ex post opportunism. To mitigate that risk, scholars reason, firms may negotiate governance mechanisms they would otherwise avoid.

First, among smaller firms, the levels of RSI are low for a simple reason: all investment levels are low. This simply is not a capital-intensive sector. Second, among larger firms (which is to say, the most productive firms), investment levels are higher—but these investments do not seem to be idiosyncratic and cross-holdings are low. These larger suppliers broadly diversify their sales outlets, and seldom issue significant equity blocks to assemblers.

Does the contracting encourage the private licensee to internalise the incentive of the owner to extract the mineral resource at the optimal rate? This may depend on the time period of the contract. The time period of the contract should be such that the public owner's optimal rate of extraction, given his/her discount rate, is the same as that of the private licensee. Long-term contracts may address the problem of contract uncertainty. Critics may argue that long-term contracts may create monopolistic situations. However, this would not apply if the licences are issued on competitive bidding and royalties are adjusted for dynamic changes over time (for example, an ad valorem royalty). Mining investments are sunk costs, irreversibly tied to a particular site and require many years to recoup. Providing security of tenure is the most critical element of practical mining law (Long, 1995).

2.2.2.2 Firm ownership and productivity

The second consideration would be whether firm ownership makes a difference to extraction efficiency or productivity. Who—SoEs or private firms—are likely to be more productive? Theoretical and empirical literature abounds on this subject.

The theory of public choice attributes the productivity difference owing to ownership to two kinds of problems: incentive and the agency problem. Public companies need not, and do not, have incentives to minimise the cost for a given level of output or maximise the output for a given level of input expenditure. This may be partly due to the fact that public-sector managers are not the claimants of the residuals from such cost minimisation. But there can also be other problems encountered by public-sector organisations such as hiring employees and fixing their wages, procurement, or inflexible budgets for investments that result in them being be less productive. A detailed account of the theoretical and empirical literature on productivity differences have been provided in Chapter 3. Overall, the studies on firm ownership and productive efficiency are inconclusive, and hence the issue remains one on which we have to rely on empirical evidence.

2.2.2.3 Profit-Inefficiency Paradox

By and large it is believed that high level of profit is associated with the high level efficiency. This may not be true in all cases. Government ownership of minerals and extractive firms may create a situation where state-owned firms generate huge (monopoly) profits but operate at low productivity frontiers. This could be due to a lack of competition. A typical example is found in the Indian coal mining industry. Coal India Limited (CIL), a state-owned enterprise, contributes more than 95% per cent of the country's total coal production. CIL records enormous profits and sits on a huge cash surplus but its productivity is not too high. Low productivity of CIL ultimately results in a failure to achieve the production targets set to match the demand for domestic consumption. This has happened in various plan periods, forcing the government to import huge quantities of coal from abroad, especially Australia. For example, in 2003, the domestic demand for coal was 382.84 million tonnes but total production was only 361.16 million tonnes, and the government had to import 21.69 million tonnes from various countries.

2.3 Equity Implications

2.3.1 Transferring surface rights and alternative compensation schemes

For extracting minerals in India, a firm needs to obtain two different rights: (i) The right to extract a mineral, which is sanctioned by the state and Central government; and (ii) the surface right to the land beneath which the mineral lies. Depending on the type of ownership, one can obtain the surface right to a particular plot of land from either the government (in the case of government land) or private tenant (in the case of private land). In the case of open-cast mining, both rights are complimentary in nature. Mineral rights sans surface right is meaningless.

In India, the land required for mining could be owned by several people as the average size of land holdings is small. There would also be some people who use such land without full ownership rights (like the tribal people using forest land). In this context, how private or public companies acquire rights to use land to extract mineral resources is a serious issue.

Theoretically, if all affected landowners have well-defined property rights, they could bargain with the mining firm—whether it be a private or public one—before transferring user rights. People could then receive a pre-agreed compensation. However, there are at least three issues that could affect the efficiency of this process. (a) Due to the large number of small landowners, the transaction cost for bargaining with all of them can be substantially high. (b) The need for RSI (in this case, the firm makes investments on the land for mineral extraction, and it may "fear"' ex-post demands for renegotiation of contracts with specific landowners) may encourage the company concerned to de-facto own lands rather than enter into a contract with each landowner. (c) If some landowners hold on to their right, without allowing the mining company to use their land to access minerals, there can be many complications. At one level, it may increase the cost of operation (if contiguous land is needed for efficient extraction). At another level, if others in the neighbourhood allow the company to extract minerals, there can be a negative externality on the piece of land which has not been surrendered.

Can the legal provisions for land acquisition be used to obtain surface rights for private mining companies? Can mineral extraction be considered a public purpose? It can be counted as a public purpose if the non-extraction of a mineral below a particular stretch of land increases social cost (by making that mineral significantly less available). The use of the Land Acquisition Act, 1894 (LAA) for a public purpose is justified when the purchase of land drastically increases the transaction cost, and when the holding-out problem makes providing for the public good costlier.

If minerals lie under the land owned by a large number of small landowners, and the government feels that extracting them will be a profitable choice, it may think of setting up a public-owned company to do this. Private parties may not show much interest in extraction because they would find it difficult to bargain with a large number of landowners to secure surface rights. But, if only a few own the land under which minerals lie, private mining firms can bargain with them to get the surface right and hence may be interested in mineral extraction.

Public companies may use the political process to acquire and compensate surface right owners. This can be part of a redistribution package—with a part of the income from mining being transferred to these people either as better compensation, or as jobs in mining companies, or through other welfare schemes. The accountability of this process is usually ensured through political means and not through legal instruments.

In a competitive market scenario, surface right owners would transfer their rights to a mining firm only when the reservation price of the former matches the latter's willingness to pay. However, in practice, the market fails to facilitate such a smooth exchange. Considering the geographical specificity of minerals and their importance to the country's economic interests, the government often intervenes to acquire land. Does the geographical specificity of minerals provide a rationale for designing a special compensation package for them? Should land or surface right owners get a share of the mineral revenue? Or should landowners in mineral-rich areas only be compensated like those in areas where other development projects are implemented? Note that a mining firm pays royalty to the owner of the mineral rights, which is the government, and other fees and taxes to it. In such a case, how can the surface right owner get a share of the profit generated from minerals which are owned by the government?

Long (1995) points out that a passive landowner has no claim to the value added in extraction and processing but can exploit the lack of alternative mining sites. Landowners want higher payments than they would get from alternative uses of the land (a reservation price), but they will negotiate for the best deal above that amount. Landowners and mining firms will negotiate respective shares of their anticipated differential rents. Long defines differential rents as the profits over and above what is necessary to bring about mineral investment. These occur when some deposits being mined are cheaper to develop and mine than others. The highest-cost profitable mining venture will generate just sufficient net income to cover the mining firms' cost of capital and the lowest of all landowners' reservation prices. Any other deposits that are cheaper to locate, develop, mine, and process will accrue differential rents.

A landowner's share of these differential rents cannot be interpreted as an in situ value of the mineral reserve since at the margin that value is zero. If all deposits were equally costly to mine, landowners would only receive the reservation price for their lands. Thus, the simple concept of being paid for the value of minerals extracted is flawed. Landowners are paid what is necessary to induce them to forego all other alternative land uses plus whatever differential rents they can bargain for (Long, 1995; p. 78).

Long suggests that an equitable and efficient method is to charge and appropriate a reservation price for surface land use, net of the value of land after reclamation, and to recover all or part of differential rents through a flat income or resource-rent tax. The traditional royalty on gross value of production, essentially a regressive income tax, cannot recover as much rent as a flat income tax, causes arbitrary mineral-reserve sterilisation, and creates a bias toward development on the extensive margin where marginal environmental costs are higher.

Thus, compensation for surface rights can be determined in alternative forms. (a) Private companies acquire surface rights through bargaining (through the market or a contract) with landowners. If the ownership of the minerals is vested in the state, and private companies have acquired mining rights through licensing, they may pay the government royalties, which in turn may come back to the people at large (not only the landowners) as public goods. (b) Alternatively, public companies providing compensation (probably better than market-based transfer of rights) and probably jobs to the landowners. This can be interpreted as a different redistribution policy—a greater part of the income from mining being shared with landowners of the mining area, rather than giving them only a part of the public goods from the public exchequer through royalties or income from mining by public-sector companies.

Thus, these different compensation policies may affect only distribution of gains between different sections in society (people living in the mining area and others within the state), and may not have much impact on efficiency, provided the high transaction cost does not reduce the aggregate surplus.

2.3.2 Differential Compensation by Public and Private Mining Firms

While discussing the compensation provided by mining firms to tenants, we can also expect a difference in behaviour between the public and private mining firms for multiple reasons. Public-sector firms, by and large, focus on maximising social welfare, even sometimes compromising their profit; whereas, private firms solely focus on maximising their profit. However, public-sector firms have the advantage of using coercive powers to acquire private land. Private firms may take the incentive route to ease land acquisition by paying higher compensation. Budgetary flexibility and government ownership of public land may allow SoEs to provide better compensation in the form of jobs, cash and land. The scale of operation and the willingness of private companies to bribe the regulatory authorities may also result in differences in the compensation packages of public and private mining firms. The causes of differences in providing compensation have been elaborately discussed in Chapter 5. All the theoretical arguments do not lead us to a single conclusion on whether public or private firms provide better compensation and the issue remains one to be decided by empirical means.

2.4. Environmental Implications

Mining may cause negative externalities, for instance, by the discharge of polluted water, suspended particulate matter and other contaminants. Such pollution may affect a large number of people and we cannot expect them to come together to negotiate with the polluter (as in Coasean bargaining). Even if they do so, negotiated outcomes may have distributional consequences. This calls for government intervention. However, government intervention need not imply the government ownership of mining companies. The government should come out with appropriate regulations and have an effective agency to enforce them.

In the first part of this section, we discuss the environmental regulations in India that have a bearing on mining operations. This includes a brief account of the failure of regulation to check environmental damage in mining areas. In the second part, we briefly discuss why the environmental performances of public and private mining firms differ.

2.4.1 Regulatory failures

For achieving sustainable development, mining operations have to safeguard the local environment and ensure that they have zero, or the least, effect on the quality of air, water, soil, and biomass. Towards this, a slew of rules and regulations have been put in place by the Central and state governments. They include the Water (Prevention and Control of Pollution) Act, 1974; the Forest (Conservation) Act, 1980; the Air (Prevention and Control of Pollution) Act, 1981; and the Environment (Protection) Act, 1986.

According to the Environmental Impact Assessment (EIA) notification dated 27 January 1994, mining of major minerals in a lease area more than 5 ha requires environmental clearance. After a Supreme Court judgment on 18 March 2004 (in the matter of Writ Petition (civil) 4677 of 1985, M. C. Mehta Vs. Union of India and Others), the EIA notification was amended on 28 October 2004 to include all mining projects in more than 5 ha that had until then not obtained environmental clearance. They were required to obtain the clearance at the time the lease came up for renewal. The environmental clearance procedure has three components. First, an EIA study has to be submitted, and there are special rules on its formulation and appraisal. Second, a public hearing has to be conducted and the procedure for this is laid down in detail. Third, an environmental management plan (EMP) has to be submitted and a separate clearance has to be obtained for it. Under the EIA, mining companies are expected to comprehensively assess all probable impact on the environment in mining areas. It also provides for ceasing mining operations in environmentally sensitive areas. To incorporate the views of local people on mining projects, the procedure stipulates holding public hearings in the mining regions. EMPs outline the strategies to be adopted by mining companies to mitigate environmental damage such as degradation of forests, and air, water and noise pollution. Under this, mining companies are supposed to undertake compensatory afforestation programmes, and check air, water and noise pollution by managing the overburden dumps and other wastes originating from mines.

Mining activities are regulated by the Pollution Control Board (for air and water), the Ministry of Environment and Forests, the Indian Bureau of Mines and state mines and geological departments. However, poor monitoring and rent-seeking activities by these agencies have combined to make the present system a failure (Agrawal, 2007; Moody, 2007). This has been demonstrated on many occasions. Starting from the EIA to adherence to the EMP, mining companies seldom comply with environmental norms. Public hearings before receiving the environmental clearance for projects are reportedly held in secret without letting the local people know (George, 2005). Compensatory afforestation programmes are confined to office documents (IT, 2008; Dharitri, 2008). Similarly, guidelines on the management of overburden dumps and other hazardous wastes have been grossly violated. Time and again local people have alleged inadequate management of mining wastes. For example, the treatment of poisonous mining waste water is carried out only irregularly, mostly when companies have received information on an impending visit by pollution regulation authorities. The violation of all environmental rules has caused serious damage to the health of local people and their environment.

The degree of negligence can be gauged from the environmental clearance allegedly accorded to the Vedanta Alumina Limited for bauxite mining in Lanjigarh in Kalahandi district of Odisha.[3] A Central Empowered Committee (CEC) that looked into the issue observed that the clearance had violated the Forest (Conservation) Act, 1980 and there had been non-payment or underpayment of compensation. There had also been gross violations of the guidelines on holding a public hearing while carrying out the EIA, the CEC reported. Very much like this, a number of other mining projects have violated the guidelines on holding public hearings.

2.4.2 Firm ownership and Environmental Compliance

Do public and private mining firms comply with environmental regulations differently? If yes, what are the reasons for the differences? Can this be due to an institutional environment in which the enforcement of environmental regulations is weak? If environmental regulations are formulated and enforced strictly, one should not see any significant difference in compliance between private and public companies. But, if enforcement is weak, there could be different patterns.

In the absence of proper enforcement of environmental standards, and the presence of a strong profit motive among private organisations, one can expect them to take advantage of the situation and emit pollution at high levels. So the absence of a profit motive in public-sector companies may be a deterrent to pollution. Strangely enough, the inefficiency of public-sector organisations may be a plus point here.

Given that the public-sector companies are owned by the citizens, it is possible to think of them internalising the cost of environmental pollution, or people exercising the demand for pollution control through the political process. However, there can be a counter-question. If people do exercise such an option, why do they not do so in the case of private mining companies as well? Otherwise, the cost of enforcing environmental regulations by a government agency must be substantially higher than the cost of self-monitoring by all firms. However, there can also be counter pressures within public firms—to decide production levels, and the like on the basis of political and other (for example, employment) considerations, marginalising environmental factors.

However, for natural resource-using industries, there can be another concern. More efficient firms are more likely to be careful in the way they

[3] See the report by the Central Empowered Committee in IA No. 1324 on the alumina refinery plant being set up by Vedanta Alumina Limited at Lanjigarh in Kalahandi district, Odisha, dated 21 September 2005. Accessed on 23 May 2008 at www.cgnet.in/Min/ CEC%20on%20Vedanta.doc.

use natural resources. Greater pollution can also be associated with inefficiency in resource use (for example, by not opting for recycling wherever it is possible). This could be a reason that could make public-sector organisations bigger polluters than private-sector firms.

Keeping in view these costs of production, differences in the environmental performance of public and private firms could emanate from three channels. First, a reason could be the strong bargaining powers SoEs have when it comes to dealing with regulatory bodies. Private firms are generally weak in this area, but rent-seeking activities by them could reverse the scenario. Second, private firms do not heed the environmental cost and use more pollution-generating inputs compared to SoEs, which are more aware of their social responsibility. Third, production efficiency results in immense differences in environmental compliance. If private firms have higher efficiencies than public firms, their environmental performance is bound to be better. Detailed accounts of the other causes of differences have been provided in Chapter 4.

2.5 Conclusion

In this chapter, we first weighed the merits of different types of ownership of minerals—individual, household, community, and different tiers of government—and came to the conclusion that the merits of state ownership outweigh those of others. Second, we analysed the implications of public and private ownership of extractive firms on four crucial dimensions—equity, efficiency, environment and revenue collection. From the analysis of ownership of minerals we concluded that state ownership would be socially efficient. Differences between public and private mining firms in productivity, environmental compliance and compensating tenants remain an empirical issue. Empirical evidence for these questions in the Indian context are provided in Chapters 3, 4 and 5.

CHAPTER THREE

ARE PRIVATE FIRMS MORE PRODUCTIVE THAN PUBLIC FIRMS?

3.1. Introduction

In this chapter we shall examine whether private firms are more productive than the public firms? Here we shall compare (i) TFP levels and growth rates of public and private firms in the four sectors of the Indian mining industry—metallic, non-metallic, coal and petroleum; and (ii) TFP levels and growth rates in the pre- and post-liberalisation periods.

This chapter has been organised as follows. Section 3.2 reviews the theoretical and empirical literature on the effect of firm ownership on productivity. Section 3.3 illustrates the methodology and data sets used for the study. Section 3.4 discusses the results of the analysis, and a summary and conclusions are presented in Section 3.5.

3.2 Firm Ownership and Productivity: Insights from Literature

3.2.1 Theoretical literature

The economic literature on firm ownership and productive efficiency is highly divided. While one section of the literature underscores the superiority of private firms over public firms (Stiglitz, 1988; Sheshinski et. al., 2003; Li and Xia, 2008, Majumdar 1998), the other section opposes this view (Caves and Christen 1980). This section presents a summary of the selected literature on the effect of firm ownership on productivity.

Ownership is seen as causing differences in firm-level productivity. This has been extensively discussed in the theories of public choice and property rights. The public choice theory attributes the productivity difference owing to ownership to two kinds of problems: incentives and the agency problem. Stiglitz (1988) attributes the inefficiency of public-sector enterprises (PSEs) to lack of two incentives: organisational and individual. Unlike the managers of private enterprise (PEs), managers of PSEs do not show much concern about the bankruptcy or the competitiveness of their companies. Due to soft budget constraints and lack of monitoring, managers of PSEs perform poorly. The losses of PSEs are very often adjusted through budgetary support. Moreover, the objective function of the PSEs is distorted by political interference (Sheshinski et. al., 2003). As the "empire building" hypothesis states, public managers incorporate their political career objectives into the objective function of PSEs, thus giving importance to employment generation at the cost of efficiency. On the individual front, predetermined pay scales and fixed tenures do not provide much incentive to increase the efficiency of the managers and workers of PSEs. The salaries of the managers and workers of PSEs are seldom linked to profit. This does not provide any incentive to put in their best effort to maximise the profits of PSEs. Similarly, fixed tenures do not encourage PSE employees to work hard.

The inefficiency of PSEs is further explained by the agency problem. The agency problem, also referred to as the principal-agent problem, in PSEs emerges due to a difference in ownership (principals) and management (agents). Here the challenge is how to ensure that the agents manage the resources in the interests of the principals instead of using their delegated authority over other people's resources to feather their own nests. PSEs suffer more from the agency problem than PEs because of the lack of a clear definition of ownership or a large number of owners—their resources belong to all the citizens of a country. Managers possess better information about a firm than its owners(s). In the absence of a proper incentive structure or monitoring by the owner(s), managers misuse their status for personal gain at the expense of the investor's interest (Li and Xia, 2008). This enables managers of PSEs to pursue wasteful projects in their own interest.

The property-right theorists' explanation for the inefficiency of public ownership comes as early as 1965. Alchian (1965) suggests that property rights are more attenuated in a public corporation than in a private corporation. The essential argument is that public ownership is diffused among all members of society, and no member has the right to sell his/her share. Given these aspects of public ownership, there is little economic incentive for any owner to monitor the behaviour of the firm's management. In contrast, it is argued that the ownership of private firms is concentrated among fewer individuals, each having the right to sell his/her shares; and thus the owners have incentives to scrutinise management to ensure production efficiency.

3.2.2 Empirical literature

A number of empirical studies have examined the performance of private- and public-sector firms. Studies that have shown the superiority of private firms over public ones are Majumdar (1998), Faria et. al. (2005) and Li and Xia (2008). Majumdar (1998) in his study of government-owned, mixed and private-sector enterprises in India from 1973–1974 to 1988–1989 shows that enterprises owned by the Central and state governments are less efficient than mixed or private-sector enterprises, while mixed enterprises are less efficient than those in the private sector. Moreover, the study shows inter-temporal efficiency gains for the sector as a whole. Faria et. al. (2005), in a study of Brazilian private and public companies providing water and sewerage utilities, show that the private companies are only marginally more efficient than the public ones. Li and Xia (2008) in a study of Chinese state firms and non-state firms establish that those in the latter category care more about efficiency than those in the former group.

Studies showing either the superiority of public firms or their neutral status are Caves and Christensen (1980), Boardman and Vining (1989) and Issac et. al. (1994). Caves and Christensen (1980) in their study of Canadian railroads shows that after controlling for the influence of regulation, there is little indication of inferior performance by the government-owned enterprise. It conveys that public ownership is not inherently less efficient than private ownership, rather the inefficiency of government enterprises stems from their isolation from effective competition rather than public ownership per se. Boardman and Vining (1989) in their literature survey show that the empirical evidence on the efficiency of public and private enterprises vary considerably across sectors. In sectors where there is some evidence of better performance by the public sector, such as electricity and water, there is limited competition or private firms are highly regulated. Evidence of the greater efficiency of PEs appears to be in the delivery of services where the government sub-contracts to the private sector. The health-related literature also suggests greater efficiency in the private sector. Moreover, Boardman and Vining argue that most of the comparative studies were carried out either for natural monopolies or regulated duopolies. Therefore, the comparison would be inappropriate.

In an attempt to explain the inconclusive result, Issac et. al. (1994) contend that the age of different enterprises may matter in discerning significant differences in productivity and efficiency. Their study shows that in the long run, productivity growth (or cost decline) is higher among private firms than state-owned enterprises. However, a comparison of productivity levels across public and private firms may produce inconsistent results depending on the age structure of the firms compared. Isaac et. al. maintain that this finding holds true in all kind of market structures, that is, whether firms operate in a competitive environment or in a more or less regulated market. In sum, the studies on ownership and productive efficiency are inconclusive and remain as empirical issue.

3.3 Methodology and Data

3.3.1 Methodology

For comparing the productivity of private- and public-sector mining firms, we first estimate the firm-level TFP and then compute the weighted average TFP levels for each sector using the share of each firm in the specific sector as weights.

Productivity is defined as the output per unit of representative units of resources (Abramovitz, 1956) and is measured as a ratio of output(s) to input(s). TFP, which is defined as the ratio of output to a combination of inputs, is being seen as a measure of the technological change that has taken place in an industry. TFP growth is defined as a residual of growth of output minus growth of input(s) or "the change in output levels after controlling for changes in input levels, or alternatively as the change in unit cost controlling for changes in input price" (Nishimizu and Page, Jr., 1986; p. 241). In the present study, TFP has been estimated as a residual in the production function.

We specify the production function for i^{th} firm in the year t as follows:

$$q_{it} = a_0 + a_c c_{it} + a_l l_{it} + a_e e_{it} + u_{it} \qquad ...(1)$$

and productivity is estimated as

$$P_{it} = q_{it} - a_c c_{it} - a_l l_{it} - a_e e_{it}$$

where q_{it} is the log of gross output in the year t, c_{it} is the log of capital stock, l_{it} is the log of labour hours, e_{it} is the log of expenditure on energy (comprises of power and fuel) and u_{it} is an error term and P_{it} is log value of firm-specific productivity. Then we take the antilog of P_{it} in order to arrive at the estimated productivity. Therefore, we first need to estimate the input coefficients from the production function which is defined in Cobb-Douglas form.

The advantages of the production-function approach (or econometric method) in estimating TFP are its liberty from the restrictive assumptions of a competitive market, constant return to scale, uniform price and technology (as taken in growth accounting method). However, it is inflicted by other problems[1] like simultaneity. Simultaneity occurs when one or more variables in the right hand side of the casual inferential model equation and the variable in the left side of the same model equation influence each other at the same time. Simultaneity causes the Ordinary Least Square (OLS)-estimated coefficients to be biased. In production function estimation it refers to the fact that at least a part of the TFP is observable by the firm at a point in time early enough so as to allow the firm to change the factor input decision. If that is the case, then profit maximisation of the firm implies that the realisation of the error term of the production function (TFP) is expected to influence the choice of factor inputs. This means that the regressors and the error term are correlated, which makes OLS estimates biased. In presence of the simultaneity, the OLS estimates will give biased results. Levinson and Petrin (2003) explain three types of biases caused due to simultaneity. (i) If only labour responds to the shock (say more labour is hired in response to a productivity shock), and capital is not correlated with labour then estimates for labour coefficient will tend to be biased up but capital efficient will remain unbiased. (ii) If only labour responds to the shock and capital and labour are positively correlated a negative bias on the capital coefficient can also result. Finally, if capital and labour are positively correlated and labour's correlation with the productivity shock is higher than capital's correlation, labour coefficient will tend to be biased upward (overestimated) and capital coefficient would be biased downward (underestimated).

---

[1] For a comprehensive review of the literature on the problems associated with the estimation of production function and the evolution of methodologies to address this problem, see Griliches and Mairesse (1998).

Symbolically, in equation (1) the error term u_{it} can be decomposed into a productivity component, that is, ω_{it} and other errors (such as estimation, errors in data and so on), that is, η_{it}. Thus, $u_{it} = \omega_{it} + \eta_{it}$. Unbiased estimation of equation (1) with the ordinary least squares (OLS) method required stochastic error terms and exogenous inputs. However, very often this condition is not met as the error term ω_{it} and inputs (mostly variable) are correlated. Marschack and Andrews (1944) point out that since inputs are chosen by a firm and not known to the econometrician, there is endogeneity in the estimation equation, which would make OLS estimates inconsistent.

A number of methodologies have evolved over time to address the simultaneity problem; the first being the within transformation. However, application of the within transformation demonstrated that either it was not doing enough, in the sense that there were still potential simultaneity problems, or was doing too much, in the sense that the transformation might be aggravating other pre-existing problems such as errors in variables. Chamberlain (1982) suggested for the first differences (or longer differences) of the available panel data, rather than going for the within transformation. If the error term gets transmitted to the current period variable inputs, then the difference in the variable input needs to be instrumented. Because the number of available instruments depends on the length of the panel and changes from one cross-section to another, optimal estimation procedures become more complex, calling for the use of generalised-method-of-moments (GMM) estimators (for example, Arellano and Bond, 1991). Griliches and Mairesse (1998) point out that such "internal" instruments (past levels for current differences and past differences for current levels) are likely to be quite poor and possess little resolving power. They suggest doing better will require bringing in additional information from somewhere else—"external" instruments, more theoretical restrictions on the structure, and/or more equations.

Olley and Pakes (1996) [OP] addresses the simultaneity problem through a proxy for the unobserved transmitted component, ω_{it}, by bringing in a new equation, the investment equation. Following Pakes (1996), Olley and Pakes assume that a firm's investment is an increasing function of productivity shocks. Which means better productivity in the present reflects better productivity in the future and this encourages capital accumulation. However, investment value for the firms making only intermittent investments will have their zero-investment observations truncated from estimation routine (the monotonicity condition does not hold for these observations). This causes the deletion of a substantial number of firms from the analysis. However, the use of intermediate inputs as a proxy alleviates this problem. Levinsohn and Petrin (2003) further refine the technique of Olley and Pakes and use intermediate input (raw materials) as a proxy to control for the simultaneity problem.

Griliches and Mairesse point out that the use of a proxy for ω_{it} has several advantages over the usual within estimators (or the more general Chamberlin and GMM-type estimators). It does not assume that ω_{it} reduces itself to a "fixed" firm effect; it leaves more identifying variance in variable and fixed inputs and hence is a less costly solution to the omitted-variable and/or simultaneity problems; and it should also be substantively more informative. Biesebroeck (2007) points out that if productivity shocks are persistent, OP is the most reliable method. Moreover, if measurement errors affect inputs and investment, OP is as effective as GMM to estimate productivity levels.

The present study uses the semi-parametric method[2] of Olley and Pakes, which is further refined by Levinsohn and Petrin and uses intermediate input, namely energy consumed as a proxy to annihilate the variations that are related to the productivity term. Our preference for using energy as a proxy is justified by the primary nature of mining industry and minimal use of raw materials.

Rewrite the production function (1) as follows,

$$q_{it} = a_0 + a_c c_{it} + a_l l_{it} + a_e e_{it} + \omega_{it} + \eta_{it} \qquad ...(2)$$

In the semi-parametric estimation a firm's energy demand function $e_{it} = e_t(\omega_{it}, c_{it})$ is assumed to be monotonically increasing in productivity, conditional on its capital stock. The inverse of the energy demand function $\omega_{it} = \omega_t(e_t, c_{it})$ depends only on the observable variables c_{it} and e_{it}. Substituting this into (2) we can derive a partial linear model as follows,

$$q_{it} = a_l l_{it} + \phi_t(e_{it}, c_{it}) + \eta_{it} \qquad ...(3)$$

[2] For an elaboration on the superiority of the semi-parametric estimation method over alternative methods like instrumental variable, see Griliches and Mairesse (1998). See also Biesebroeck (2007) for a comparison of five widely used techniques to estimate TFP.

where,

$$\phi_t(e_{it}, c_{it}) = a_0 + a_e e_{it} + a_c c_{it} + \omega_t(e_{it}, c_{it})$$

The estimation of equation (3) is carried out in two stages. In the first stage, as the error term in equation (3) η_{it} is not correlated with the input *e*, we estimate the coefficient for labour (a_l) by including ϕ_t (.) in the estimation routine. In this, ϕ_t (.) is approximated by a third order polynomial with a full set of interactions. It is also allowed to be different in the three sub-periods of (in terms of policy changes) the sample, 1988-89 to 1992-93, 1993-94 to 1999-2000 and 2000-01 to 2005-06, corresponding to three distinct phases of the mining industry. Since capital and energy enters ϕ_t (.) in two ways, a complete model is used to identify the a_c and a_e.

In the second stage, ω_{it} is assumed to follow the first order Markov process. $\omega_{it} = E[\omega_{it} | \omega_{it-1}] + \xi_{it}$, where ξ_{it} is the innovation in productivity over the last period's expectation. Two moment conditions are used to identify a_c and a_e. The first moment condition to identify a_c, which assumes that capital does not respond to ξ_{it}, is

$$E[(\xi_{it} + \eta_{it})c_{it}] = E[\xi_{it} c_{it}] = 0 \qquad ...(4)$$

The second moment condition to identify a_e, assumes that the last period's energy choice is uncorrelated with ξ_{it}, and is

$$E[(\xi_{it} + \eta_{it})e_{it-1}] = E[\xi_{it} e_{it-1}] = 0 \qquad ...(5)$$

The residuals used in the moment conditions in (4) and (5) are given by

$$\xi_{it} \hat{+} \eta_{it}(a^*) = q_{it} - \hat{a}_l l_{it} - a_e^* e_{it} - a_c^* c_{it} - E[\omega_{it} \,\hat{|}\, \omega_{it-1}] \qquad ...(6)$$

where the residuals are expressed as a function of the two parameters $a^* = (a_e^*, a_e^*)$. $E[\omega_{it} \mid \omega_{it-1}]$ is estimated by regressing $\hat{\omega}_{it}$ on fourth order polynomials in $\hat{\omega}_{it-1}$. $\hat{\omega}_{it}$ and $\hat{\omega}_{it-1}$ are respectively obtained from the following equations (7) and (8) using the estimates from the first stage (labour coefficient) and candidate values for (a_e^*, a_e^*). The candidate values are the OLS estimates of production function (1).

$$\omega_{it} \hat{+} \eta_{it} = y_{it} - \hat{a}_l l_{it} - a_c^* c_{it} - a_e^* e_{it} \quad (7)$$

$$\hat{\omega}_{it-1} = \hat{\phi}_{t-1}(.) - a_c^* c_{it-1} - a_e^* e_{it-1} \quad (8)$$

We also use the following three additional moment restrictions to test the unbiased nature of the estimated coefficients of the choice variables of a firm, namely *c, l, and, e.*

$$E\lfloor \xi_{it} l_{it-1} \rfloor = 0, \; E\lfloor \xi_{it} c_{it-1} \rfloor = 0, \; E\lfloor \xi_{it} e_{it-2} \rfloor = 0$$

These over-identifying moment restrictions are valid under the null-hypothesis that the coefficient estimates are unbiased. Thus, we have total five population moment conditions given by the vector of expectations,

$$E[(\xi_{it} + \eta_{it}) Z_{it}] = 0$$

where Z_{it} is the vector given by

$$Z_{it} = \{c_{it}, e_{it-1}, c_{it-1}, e_{it-2}, l_{it-1}\}$$

Finally, we obtain estimates $(\hat{a}_c, \hat{a}_e)$ by minimising the GMM criterion function.

$$Q(a^*) = \min \; a^* \sum_{h=1}^{5} \left(\sum_i \sum_{t=Ti\,0}^{Ti\,1} (\xi_{i,t} \hat{+} \eta_{i,t}(a^*)) Z_{i,h,t} \right)^2$$

where i and h index the firms and five instruments respectively; T_{i0} and T_{i1} index the second and last period firms i is observed.

To assess the precision of estimates, we need to take account of the variance and co-variances in every estimator that enters the estimation routine. Instead of undertaking this difficult exercise, we go for bootstrapping. This technique (re)samples the estimates from sample data to construct new "bootstrapped" samples. The value of the statistics is computed for each of these samples and the distribution of estimates so generated provides the bootstrap approximation to the true sampling distribution of the statistics. We use block bootstrapping for this purpose. In this method, the sampling rule treats each set of firm-level observations as an independent block and identical draw from the population of firms. Sampling is done with replacement and with equal probability from the sets of firm-level observations in the original sample. The number of replication is five hundred. Since the estimation includes over-identifying moment restrictions, we use recentred moments in the bootstrap estimation to make sure that bootstrap samples implement moment conditions that are valid in the original data set from which it samples (see Horowitz, 2001).

Using the estimated production function coefficients, TFP (P_{it}) levels for each firm are estimated as follows,

$$P_{it} = \exp(\ln q_{it} - \ln c \times a_c - \ln l \times a_l - \ln e \times a_e)$$

Industry-level TFP has been estimated as the weighted mean of firm-level TFP estimates where each firm's output share in the industry is used as weights.[3] In the similar fashion, weighted average TFP levels for the four sectors of the mining industry—metallic, non-metallic, coal and petroleum – has been estimated, where weights are estimated as each firm's share in the specific sector's output. Further, by dividing each sector into two ownership groups (public and private), TFP has been estimated for eight sub-sectors. For grouping the firms under the four sectors, we follow the National Industrial Classification (NIC) codes at two-digit level.[4] However, due to the presence of very few firms in category 12 (mining of thorium and uranium), those firms have been

[3] The rationale behind using the weighted mean is to reflect a better picture than simple average estimates, which give biased estimates depending on the distribution; upwardly biased in case of positively skewed and downwardly biased in case of negatively skewed.

[4] Two-digit NIC codes for the various sectors of the mining industry are as follows: 10–Mining of coal and lignite, extraction of peat; 11–Extraction of crude petroleum and natural gas, service activities incidental to oil and gas extraction including surveying; 12–mining of thorium and uranium ores; 13–mining of metal ores; and 14–Other mining and quarrying.

clubbed with those in the coal sector. Similarly, for ownership classification, all the government–Central as well as state–enterprises are under the public sector and the rest under the private sector.

3.3.2 Data

The study uses firm-level data provided by the Centre for Monitoring Indian Economy (CMIE) in its electronic data base, Prowess. Firm-level data are taken for mining firms under two-digit national industrial classification (NIC) from 1988-89 to 2005-06. Prowess is a database of large and medium Indian firms and comprises of all companies traded on India's major stock exchanges and several others including the central public sector enterprises. Prowess provides the normalised database of the financials covering 1,500 data items and ratios per company. Besides, it provides quantitative information on production, sales, consumption of raw material and energy. The reporting of firms to CMIE is purely voluntary. Therefore, the entry and exit of the firm from the sample do not reflect the true entry and exit from the industry. Hence the entry and exit is not correlated with firm's productivity performance. Therefore, the data set does not suffer from a selectivity bias (which arises due to the selection of the efficient firms present in the industry). Firms having data for at least 3 years in sequence were included for the analysis. Thus we get a sample of firms consisting of 872 observations on 65 firms. Thus the data set represents around 40% of the total value of the mining industry's output in 1988-89 and a much higher percentage in successive years. In 2003-04, the data set represents around 90% of the total value of the mining industry's output (see Table A3.2 in the Appendix). The summary measures of the variables are presented in Table A3.3 in the Appendix. The data set used for analysis is an unbalanced panel.

3.3.3 Construction of Variables

The time series variables have been measured in 1993-94 prices.[5] For this purpose, price indices are collected from the "Index Numbers of Wholesale Prices in India, base 1993-94 = 100" published by the Economic Adviser, Ministry of Commerce and Industry, Government of

[5] Bartelsman and Doms (2000) point out that using deflated production to measure productivity has one drawback. Any quality improvement in output that is not reflected in the deflator will result in a downward bias in productivity. However, given the nature of data, the present study does not do such an adjustment

India. The detailed procedures for construction of variables are explained below.

Output (*q*): The value of the output series for each firm is obtained by deflating the current year values. To minimise the discrepancy between industry-level and firm-level price deflators, disaggregated industry-level price indices have been used for the four sectors—metallic, other minerals, coal and mineral oils[6].

Capital (*c*): The sample reports firms' gross fixed asset (GFA) and its various components at historical cost. Data on Capital stock have been computed using the perpetual inventory method. To meet this objective, all given GFA values are re-valued at the replacement cost with 2002-03 as the base year. For estimating the revaluation factor, we follow the method used by Srivastava (1996).[7] This study uses the GFA rather than net fixed asset.[8] For a detailed methodology on capital stock estimation (GFA), see Appendix A.3.1.

Labour (*l*): Labour input is measured in terms of labour hours. This has been computed by dividing the total wage bill (salary and wages) given in the data base by the average wage per hour. The data for the latter is estimated from the average weekly wages in the coal and non-coal sectors published in "Statistics of Mines in India" by the Directorate General of Mines Safety, Ministry of Labour and Employment, Government of India. The per hour wage is computed on the basis of six working days in a week and eight hours of work a day.

Energy (*e*): The energy variable is constructed by deflating the reported energy (power and fuel) cost by an energy price index constructed using weights obtained from the "Input-Output Transaction Table of India for 1993-94" and appropriate price indices.

---

[6] A single deflator is more appropriate in a perfectly competitive market environment, where the law of uniform price prevails and all firms face the same price. In the present analysis, there is always a possibility of discrepancy between the firm-level and industry-level price deflators. Given data constraints, this issue remains unaddressed here.

[7] For another method, see Parameswaran (2004) and (2007).

[8] For estimating the revaluation factor of the net fixed asset of a firm, we need information on accounting and the economic rate of depreciation. Reliable data on accounting and the economic rate of depreciation are not available in India. The depreciation values reported by firms are mainly determined by the provisions of the income tax rules than by actual depreciation of assets. Further, Dennison (1967) argues that the correct measure of capital stock falls somewhere between gross and net stock of capital, advocating the use of a weighted average of the two, with higher weight for the gross asset as the true value is expected to be closer to it.

3.4 Results

This section presents the results in three parts. At the outset, production function estimates of the OLS method and Semi-Parametric Method (SPM) are reported. Next, a comparison is made between the TFP levels of public- and private-sector firms (within the sector), followed by the TFP growth rates in the pre- and post-liberalisation periods in the four sectors of the mining industry. The four sectors of the mining industry have very different pricing mechanism. For example, the price of coal and crude petrol are officially determined by the central government (administered pricing mechanism) and therefore does not reflect the true market price. But the price of metallic and non-metallic sector is by and large determined by the market forces. Productivity estimates for the sectors where administered price is prevailing would definitely show a downward bias compared to the sectors where price is market determined. In such a scenario it would be misleading to compare the productivity estimates across sectors. Nonetheless, intra-sector comparison of productivity would not cause any such bias for all the firms within the sector face same pricing scheme. Therefore, we refrain from comparing TFP levels across sectors and compare only growth rates. Finally, the productivity gap between public and private firms is tested in a regression model by controlling for initial productivity levels and the age of firms.

Production function estimates from two methodologies—Semi-Parametric Method (SPM) and OLS—are reported in Table 3.1. The results of the two estimates reveal that the coefficients of labour and capital obtained from the OLS method are biased downward and that of energy is biased upward when compared to SPM estimates. For the SPM estimates, row P(Q) reports the P value of the over-identification test under the null hypothesis that over-identification restrictions are valid. The null hypothesis is valid if the proxy is conditioning out all the variation in inputs that are correlated with unobserved productivity. Table 3.1 demonstrates that the null is accepted at a reasonable level of statistical significance.

Table-3.1: Production Function Estimates

Inputs	SPM	OLS
Labour	0.402* (0.099)	0.362* (0.028)
Capital	0.326* (0.069)	0.304* (0.026)
Energy	0.171* (0.068)	0.199* (0.023)
P(Q)	0.618	
No of Observations	872	872

Notes:
For the semi-parametric analysis, bootstrap standard errors of estimates are given in parentheses. Number of replication is 500.
For the OLS analysis, ordinary standard errors of estimates are given in parentheses.
P(Q) is the P value of the over-identification test.
* Conveys the significance of variables at 1% of confidence interval.

3.4.1 Comparison of TFP between Public and Private Firms

In this section, we compare the TFP levels of public and private firms in the four sectors of Indian mining industry—metallic, non-metallic, coal and petroleum. Table 3.2 reports the TFP levels in index form for public and private firms in the four sectors. In each sector, TFP indices of private firms are compared with those of firms in the public sector.

Table-3.2 Indices of TFP Levels for Public and Private firms in the four Sectors of Indian Mining Industry from 1988-89 to 2005-06

Year	Metallic		Non-metallic		Coal		Petroleum	
	Public	Private	Public	Private	Public	Private	Public	Private
1988-89	100.0	171.9					100.0	318.2
1989-90	126.7	199.4	100.0	203.5			153.1	466.4
1990-91	137.4	196.2	95.1	179.8	100.0	137.3	146.2	426.2
1991-92	152.8	203.8	100.3	208.0	100.7	143.1	144.8	427.3
1992-93	138.8	205.8	115.7	197.2	102.2	153.4	158.7	315.4
1993-94	141.7	199.7	145.6	227.2	105.7	218.7	171.0	292.0
1994-95	158.8	220.3	164.8	223.7	110.0	272.8	207.0	287.8
1995-96	167.5	198.0	126.5	218.8	108.7	169.5	180.8	199.3
1996-97	174.8	201.4	124.4	227.9	108.7	150.8	190.2	209.1
1997-98	160.9	205.8	127.2	231.0	113.9	148.8	195.5	194.1
1998-99	158.0	202.0	119.9	218.8	112.0	142.5	218.2	272.4
1999-00	158.0	215.4	139.0	224.0	108.9	154.5	176.2	194.1
2000-01	169.9	212.5	146.3	239.4	102.4	160.3	196.5	187.1
2001-02	176.5	221.4	162.7	236.2	110.2	153.8	223.8	175.9
2002-03	179.1	234.2	165.2	259.2	111.1	162.7	194.8	194.1
2003-04	149.3	222.0	158.5	275.6	107.6	160.6	179.0	210.8
2004-05	158.0	236.5	166.2	311.8	110.0	164.1	235.3	265.0
2005-06	151.3	220.9	172.1	322.6	117.2	167.8	238.1	222.7

Note: The TFP indices have been constructed by taking the initial year TFP value of public sector as 100. This serves as the reference value and TFP values of latter years and private sector are compared against this reference value.

During the entire period of analysis, three sectors—metallic, non-metallic and coal—recorded higher TFP levels for private firms than for public firms. However, the petroleum mining industry presented a slightly different picture. In the first phase, from 1988-89 to 1996-97, TFP levels of private firms remained consistently higher than those of their public counterparts. In the second phase, from 1997-98 to 2005-06, TFP levels of public firms outweighed their private counterparts, barring four years—1998-99, 1999-2000, 2003-04 and 2004-05.

Table-3.3 Ratio of TFP Levels of Private to that of Public Firms in the Four Sectors of the Mining Industry from 1988-89 to 2005-06

Year	Metallic	Non-metallic	Coal	Petroleum
1988-89	1.72	-	-	3.18
1989-90	1.57	2.04	-	3.04
1990-91	1.43	1.89	1.37	2.92
1991-92	1.33	2.08	1.42	2.95
1992-93	1.48	1.71	1.50	1.99
1993-94	1.41	1.56	2.07	1.71
1994-95	1.39	1.36	2.48	1.39
1995-96	1.18	1.73	1.56	1.10
1996-97	1.15	1.83	1.39	1.10
1997-98	1.28	1.81	1.31	0.99
1998-99	1.28	1.83	1.27	1.25
1999-00	1.36	1.61	1.42	1.10
2000-01	1.25	1.64	1.57	0.95
2001-02	1.25	1.45	1.39	0.79
2002-03	1.31	1.57	1.47	1.00
2003-04	1.49	1.74	1.49	1.18
2004-05	1.50	1.88	1.49	1.13
2005-06	1.46	1.87	1.43	0.94

Table 3.3 shows the ratios of TFP levels of private firms to those of public firms. Values greater than one indicate the multiples by which private firms are more productive than public firms; values less than one indicate the opposite; and values equal to one indicate zero difference. The productivity gap between public and private firms was highest in the non-metallic mining sector. Private firms in this sector were almost two times more productive than their public counterparts. Private firms in the metallic and coal mining sectors were one and half times more productive than their public counterparts. Only in the petroleum sector did TFP levels of public firms come close to or exceed that of private firms. However, a part of this can be attributed to the fall in TFP levels of private firms. Although public firms in the petroleum sector experienced a steady rise in TFP levels, private firms experienced a sharp fall. The productivity gap in the other three sectors, however, persisted steadily.

The opening up of the mining industry to greater private participation is expected to pose a threat of survival to public-sector firms. This could force public firms to enhance their productivity growth to catch up with private-sector firms. In the following paragraph, we compare the TFP growth rates of public and private firms in the four mining sectors. Table 3.4 reports the average annual TFP growth rates of private and public firms in the four sectors of Indian mining industry between 1989-90 and 2005-06.

Table-3.4: Average Annual TFP Growth Rates (in %)

Periods	Metallic		Non-metallic		Coal		Petroleum	
	Public	Private	Public	Private	Public	Private	Public	Private
1989-90 to 1992-93	9.27	4.81	5.29	-0.29	1.86	5.75	14.26	3.00
1993-94 to 1999-00	2.05	0.86	3.82	2.02	0.96	3.04	2.37	-4.31
2000-01 to 2005-06	-0.38	0.55	3.73	6.36	1.36	1.43	6.14	3.20
1989-90 to 2005-06	2.89	1.68	4.06	3.01	1.32	2.76	6.50	0.06

Note:
Keeping in view the different phases of liberalisation of the mining industry we have divided the entire period of analysis (1989-90 to 2005-06) into four sub-periods as follows:
(i) Pre-liberalisation period: 1989-90 to 1992-93:
(ii) First phase of liberalisation allowing domestic private firms and partial participation of foreign firms: 1993-94 to 1999-00,
(iii) Second phase of liberalisation allowing more foreign participation: 2000-01 to 2005-06.

Over the entire period of analysis, TFP growth of public firms was higher than that of private firms in all sectors other than coal. Nevertheless, the growth rates in three sub-periods show a somewhat different picture. During the first two sub-periods periods (from 1989-90 to 1992-93, and from 1993-94 to 1999-2000), TFP growth of public firms was higher than their private counterparts in all sectors except coal. But the trend in TFP growth was just the reverse in the last sub-period (from 2000-01 to 2005-06). During this period, TFP growth of private firms in the metallic, non-metallic and coal sectors exceeded that of public firms. However, the situation in the petroleum mining sector was the opposite. It reflected that public firms in the petroleum sector have consistently improved their productivity compared to private firms. In contrast, public firms in the coal industry have consistently had poor productivity growth.

3.4.2 TFP Growth in the Indian Mining Industry

Having discussed TFP levels and TFP growth of public and private mining firms in the industry's four sectors, we analyse TFP growth of the overall mining industry and its four sectors. Table 3.5 reports the average annual TFP growth for the mining industry and its four sectors between 1988-89 and 2005-06.

Table-3.5 Average Annual TFP (%) Growth Rates in the Four Sectors of the Indian Mining Industry

Periods	Total Mining Industry	Metallic	Non-Metallic	Coal	Petroleum
1989-90 to 1992-93	4.49	6.77	-1.25	2.23	18.10
1993-94 to 1999-00	0.94	1.50	2.46	1.04	-3.60
2000-01 to 2005-06	3.04	0.44	6.29	1.43	5.26
1993-94 to 2005-06	1.91	1.01	4.23	1.22	0.49
1989-90 to 2005-06	2.52	2.37	2.94	1.46	4.63

The mining industry's TFP level has increased by almost one and half times from 4.37 in 1988-89 to 6.55 in 2005-06 (see Table A3.4 in Appendix). During the entire period of analysis, TFP of the aggregate mining industry grew at the rate of 2.52% per annum. The pre-liberalisation period (1988-89 to 1992-93) recorded the highest rate of TFP growth, 4.49% per annum. The first phase of the post-liberalisation period (1993-94 to 1999-2000) saw a marked decline in TFP growth with it being 0.97% per annum. The second phase of post-liberalisation (2000-01 to 2005-06) saw a recovery; nonetheless the growth could not bounce back to the level of the pre-liberalisation period. TFP growth during this period was recorded at 3.04% per annum. Thus average annual TFP growth in the post-liberalisation period has been lower (1.91% per annum) than in the pre-liberalisation period.

Sectoral analysis reveals a very different picture in each sector. In the post-liberalisation period, TFP growth in the metallic, coal and petroleum sectors declined compared to the pre-liberalisation period. Only the non-metallic sector showed a steady rise in TFP growth in the post-liberalisation period. After recording average annual negative TFP growth (-1.25% per annum) in the pre-liberalisation period, the non-metallic sector experienced a steady rise in TFP growth in the post-liberalisation period (4.23% per annum). Like the aggregate mining industry, the coal and petroleum sectors experienced a slowdown in TFP growth in the first phase of liberalisation but recovered in the second phase of liberalisation. The metallic sector, however, showed a steady decline in TFP growth in the post-liberalisation period. Thus, over the entire period of analysis, the petroleum sector recorded the highest rate of TFP growth (4.63% per annum), followed by the non-metallic (2.94% per annum), metallic (2.37 % per annum) and coal (1.46% per annum) sectors. However, in the post-liberalisation period, TFP growth in the non-metallic sector was the highest at 4.23% per annum and it was the lowest in the metallic sector at 1.01% per annum.

Although the above discussion clearly shows the productivity gap between public and private firms, we go a step further to confirm this by controlling for other factors such as the age and initial productivity level of a firm. Here it is worth noting that the age of firms might cause differences in the productivity of public and private firms (Issac et. al., 1994). Therefore, we specify a multivariate regression model with an unweighted TFP level as the dependent variable, and lagged values of TFP level, age and ownership of the firm as independent variables. Inclusion of a lagged dependent variable on the right hand side of the equation makes the model autoregressive, which is specified as follows,

$$P_{it} = \alpha_p P_{i,t-1} + \alpha_{a1} a_{it} + \alpha_o O_i + \eta_i + v_{i,t}$$

for i = 1..,N and t=2,..,T, where P_i is log values of TFP levels of the firm i in period t, O is dummy variable for firm ownership (Public =0, Private =1), a_{it} is log value of age of the firm i in period t, $u_{it} = \eta_i + v_{it}$ is the usual "fixed effects" decomposition of the error term. To address the endogeneity problem in an autoregressive model, we adopt the combined estimation, GMM-system, a Blundell and Bond (1998) method, which uses lagged differences of P_{it} as instruments for equations in levels in addition to lagged (from second to fifth) levels of P_{it} for equation in differences. The above equation is estimated for the mining industry as a whole and three of its sectors (metallic, coal and petroleum) separately by the software DPD for Ox (Doornik et. al., 2001; 2008).

Table-3.6: Results of Dynamic Panel Regression Analysis

Independent Variables	Model-1 Total Industry	Model-2 Metallic	Model-4 Coal	Model-5 Petroleum
Dependent Variable: P_{it}				
$P_{i,(t-1)}$	0.427**	0.21**	0.27**	-0.08
	(0.099)	(0.012)	(0.02)	(0.28)
$a_{i,t}$	0.005	0.34**	0.31**	0.35**
	0.04	(0.009)	(0.013)	(0.06)
O_i	0.11*	0.15**	0.14**	0.66
	(0.05)	(0.03)	(0.05)	(0.29)
Constant	0.9**	0.104**	0.16**	0.3**
	(0.18)	(0.001)	(0.01)	(0.1)
Sargan	62.37	0.000	0.000	0.000
	[0.784]	[1.000]	[1.000]	[1.000]
AR(2)	-1.116	0.94	-1.078	-0.87
	[0.264]	[0.347]	[0.281]	[0.384]
Number of Observations	802	115	210	101

Notes:
Standard errors of the two-step GMM estimators are corrected for small sample bias as in Windmeijer (2006) and presented in parentheses.
Sargan test P-values are in parentheses.
*, ** indicate significance at 5% and 1% level respectively.

Sectoral analysis enables us to avoid a faulty representation of results from pooled analysis, ignoring the differential share of each sector. Private mining firms are hypothesised to be more productive than their public counterparts. Hence the ownership dummy is expected to have a positive sign.

Results of the regression analysis are presented in Table 3.6. Note that the Sargan-test statistics signal the validity of the over-identifying restrictions, which in turn validates the instruments used. Further, absence of second-order serial correlation of the errors—indicated by statistic labelled AR (2)—satisfies the specific requirement for instrument validity when the second lag has been included in the instrument set, as has been done here. The results indicate that TFP of private firms in the mining industry is significantly higher than that of public firms. A sector-wise analysis also confirms such a difference in the metallic, coal and petroleum sectors. Similarly, lagged TFP levels have a positive and significant influence on present TFP levels of the mining industry and in the sectoral analysis, this applies to the metallic and coal sectors.

However, this does not hold true for the petroleum sector. At the industry level, the age of a firm does not seem to have a significant effect on TFP levels. However, the sectoral analysis shows a positive relationship between the age of a firm and TFP levels in the metallic, coal and petroleum sectors.

3.5 Conclusion

In this chapter we sought to compare the productivity difference between public and private mining firms in the four sectors in pre- and post-liberalisation periods. The comparison of TFP levels of public and private firms between 1988-89 and 2005-06 shows the superiority of private companies in three sectors—metallic, non-metallic and coal—during the period of analysis. In the petroleum sector, private firms initially outperformed public firms but eventually TFP levels of public firms exceeded those of private firms in a few years. The productivity gap between public and private firms was highest in the non-metallic sector. Private firms in this sector were almost two times more productive than their public counterparts. Private firms in the metallic and coal sectors were one and half times more productive than their public counterparts. The competitive pressure brought in by liberalisation has not been able to bridge the productivity gap between public and private firms.

Liberalisation of the mining industry has resulted in an increased private participation. Consequently, the share of the public sector in the total value of output has declined from 91.19% in 1988-89 to 74.61% in 2004-05. Private participation has been the highest in the non-metallic sector. For example, in 2003-04, the share of the private sector in the value of total limestone production was 93.6%. Between 1988-89 and 2005-06, the mining industry's TFP rose by one and half times and on an average grew at the rate of 2.52% per annum. However, inter-temporal comparison of TFP growth shows differential performances in three sub-periods. In the pre-liberalisation period, TFP of the mining industry grew at a higher rate than in the post-liberalisation period. While TFP in the pre-liberalisation period grew at around 4.5% per annum, it experienced a slowdown in the first phase of liberalisation and declined to 0.94% per annum. However, the industry bounced back in the second phase of liberalisation and TFP grew at around 3% per annum. On average, TFP growth in the post-liberalisation period (1.91%) has been lower than that of the pre-liberalisation period (4.49%).

Appendices

A3.1 Measurement of Capital

Measuring the contribution of capital to total output has been a very contentious issue in economics and statistics. This was reflected in the 18th meeting of the UN Statistical Commission's Working Group on International Statistical Programmes and Co-ordination, which met in April 1996, including capital stock measurement in its list of 15 so-called critical problems in economic statistics. This was also considered to be one of the most difficult tasks that economists had set for statisticians by the Statistical Division of the UN Economic Commission for Europe (UNECESD, 2003; p. viii). Apart from confusion on the factors to be enlisted under the category of capital, other crucial issues such as utilisation of capacity, depreciation and the obsolescence of capital draw vital attention. Yet, there is now wide agreement that the contribution of capital to production should be measured in terms of the "flow of services" produced by capital assets rather than by a "stock" of those assets (OECD, 2001). However, reliable data on the flow of services produced by capital assets in India is not available, and we are constrained to employ data on capital stock. For measuring the capital stock of a firm, this study follows the method used by Srivastava (1996), which revalues the capital given at historical cost to a base year. The detailed computing routine is presented below.

The Prowess data base provides the information on GFA at historical cost, its three components (land, building and plant), machinery and depreciation. Actual investment in a year is estimated by calculating the difference in GFA between the current year and the previous year. The nominal values for investment have been converted to real values in the 1993-94 base price. This enables us to use the perpetual inventory method to construct capital stock as shown below.

$$k_{t+1} = k_t + I_{t+1}$$

$$k_t = k_{t-1} + I_t$$

$$k_{t-2} = k_t - I_t - I_{t-1}$$

and so on. Where, k_{t+p} and I_{t+p} are the capital stock and the real investment respectively for the period $t + p$.

To arrive at this estimation, we first need to revalue the GFA at historical cost to a particular base year value. The rationale behind the revaluation of capital to a base year price is worth noting. To derive the share of capital and labour from the estimation of production function, the value of all parameters (output and inputs) should be expressed in one base year. Given the output and variable input value at current price, the cost of capital also needs to be expressed in terms of the replacement cost in the same period. However, a separate revaluation of capital for each firm would mean a Herculean task. Therefore, we resort to a number of simplifying assumptions. First, we estimate a revaluation factor to convert the historical GFA to replacement cost in a base year. To ease the task, we apply the same revaluation factor to firms employing capital of similar vintage. For estimating the revaluation factor, we choose a base year having the maximum number of firms. Thus, in the present case, 2002-03 has been selected as the base year. Before moving ahead, one needs to understand that the capital employed by a firm operates only for a specific period of time. So, depending on the year of incorporation, machines will function only for a fixed duration. On this account, we take the life tenure of capitals employed in the mining sector published by in the "National Accounts Statistics—Sources and Methods, 2007" by the Central Statistical Organisation (CSO[9]), New Delhi.

The estimation of the revaluation factor involves the following assumptions.

1. Given 25 years of life for capital and the base year 2002-03, it is presumed that the no firm has vintage capital earlier than 1978-79 and firms incorporated in that year have employed vintage capital of the same year. Thus firms incorporated (after 1978-79) in the same years are presumed to have the same vintage capital and firms incorporated before 1978-79 to have the vintage capital of 1978-79.
2. The price of capital is assumed to have changed at a uniform rate $(\prod = \frac{P_t}{P_{t-1}} - 1)$ from 1978-79 or the year of incorporation, whichever is later, to 2002-03 for all firms. Values for $\prod$ are

[9] For estimating the life of capital employed in various sectors, the CSO considers the data on the average life of machine tools in the reports of Censuses of Machine Tools (1968 and 1986) conducted by the Central Machine Tools Institute, Bangalore, and applies the depreciation provision under income-tax rules as well as the Companies (Amendment) Act, 1988.

estimated by constructing a price index for the gross capital formation for the mining and quarrying sector, compiling data from various volumes of National Account Statistics of India.

3. Similarly, investment is assumed to have changed at a uniform rate ($g = \frac{I_t}{I_{t-1}}$) from 1978-79 to 2002-03 for firms incorporated during the period. Here the growth rate of gross fixed capital formation in the manufacturing sector at 1999-2000 prices is assumed to apply to all firms. Further, average annual growth rates are obtained for firms established after 1978-79.

With these underlying assumptions we derive a series of revaluation factor (R_t^G) for the years between 1978-79 and 2002-03. The revaluation factor is multiplied to the value of capital at historical cost (VCHC) to derive the value of capital stock at replacement cost (VCRC). Thus VCRC $= R_t^G \times$ [VCHC]

A series of revaluation factors is derived as explained below:

A3.1.1: Revaluation Factor for Gross Fixed Assets

Let us denote GFA_t^H and GFA_t^R as GFA at historical costs and replacement costs respectively and I_t is the real investment at time *t*. By definition, as well as following the above mentioned assumptions, the equation for GFA_t^H can be written as follows,

$$GFA_t^H = P_t I_t + P_{t-1} I_{t-1} + P_{t-2} I_{t-2} + \ldots$$

$$= P_t I_t \left(\frac{(1+g)(1+\Pi)}{(1+g)(1+\Pi)-1} \right)$$

and $GFA_t^R = P_t I_t + P_t I_{t-1} + P_t I_{t-2} + \ldots$

$$= P_t I_t \left(\frac{(1+g)}{g} \right)$$

Defining $R_t^G = \dfrac{GFA_t^R}{GFA_t^H}$

Then $R_t^G = \dfrac{(1+g)(1+\Pi)-1}{g\,(1+\Pi)}$

If it is assumed more realistically that the capital stock does not date back infinitely, but the capital stock of the earliest vintage is *t* period old, then we can derive the revaluation factor as follows,

$$R_t^G = \frac{\left[(1+g)^{t+1}-1\right](1+\Pi)^t\left[(1+g)(1+\Pi)-1\right]}{g\left\{\left[(1+g)(1+\Pi)\right]^{t+1}-1\right\}}$$

The GFA thus obtained is used, after deflating it with the wholesale price index for machinery and machine tools, as plant and machinery account for 71% of the GFA (RBI Bulletin, 1990). Finally, in this study, the GFA of firms is used rather than net fixed asset.

Table- A3.1: Sector and Year wise Sample Number of Firms in Indian Mining Industry

Year	Mining Industry	Metallic		Non-Metallic		Coal		Petroleum	
		Public	Private	Public	Private	Public	Private	Public	Private
1988-89	16	3	2	0	1	8	0	1	1
1989-90	18	3	2	1	2	8	0	1	1
1990-91	23	3	2	1	6	8	1	1	1
1991-92	25	3	2	1	7	8	2	1	1
1992-93	36	3	2	1	13	10	2	1	4
1993-94	42	4	3	1	17	10	2	1	4
1994-95	48	4	3	1	22	10	2	2	4
1995-96	55	4	3	1	26	10	4	2	5
1996-97	59	4	3	1	30	10	4	2	5
1997-98	59	4	3	1	30	10	4	2	4
1998-99	60	4	3	1	31	10	4	2	3
1999-00	63	4	4	1	32	10	3	3	5
2000-01	62	4	4	1	31	10	4	3	5
2001-02	62	4	4	1	30	10	4	4	5
2002-03	62	4	4	1	30	10	4	4	5
2003-04	62	4	4	1	30	10	4	4	5
2004-05	61	4	4	1	29	10	4	4	5
2005-06	59	4	4	1	27	10	4	4	5

Source: CMIE

Table A3.2: Representation of Prowess Output as a Share of Total Output in the Mining Industry

Year	Total Value of Output of Mining Industry Rs. in Crore	Total Value represented by CMIE Data Rs. in Crore	% Representation by Prowess
1988-89	17913	7177.13	40.07
1989-90	19123	8026.94	41.97
1990-91	20754	8760.878	42.21
1991-92	23284	10529.09	45.22
1992-93	27040	13279.55	49.11
1993-94	30736	15694.83	51.06
1994-95	33998	25676.24	75.52
1995-96	38269	27024.43	70.62
1996-97	44193	33076.48	74.85
1997-98	45418	38608.66	85.01
1998-99	52307	36960.4	70.66
1999-00	58765	45469.73	77.38
2000-01	60931	51304.96	84.20
2001-02	66878	53253.07	79.63
2002-03	71382	64767.89	90.73
2003-04	75018	67839.64	90.43
2004-05	75128	91563.31	
2005-06	75239	102085.2	

Note: Computation of the value of total output for the Indian mining industry remains a daunting task due to unavailability of systematic data. Different data sources provide different estimates. The data used in this table are taken from various issues of the Indian Bureau of Mines (IBM) publication "Indian Mineral Industry at a Glance". This estimate excludes the value of minerals declared as prescribed substances under the Atomic Energy Act, 1962. However, the CMIE sample data includes these values. Therefore, in the last two years, the value of output estimated from CMIE exceeds the total value of mineral production given by IBM. The data published by the CSO in its "Annual Survey of Industries" provides only a part of the total value of mineral production.

Table-A.3.3: Summary Statistics of the Data Used for Analysis

Year	Real output value in crore		Labour hours		Capital stock in Rs. Crore		Fuel & energy in Rs. Crore	
	Average	SD	Average	SD	Average	SD	Average	SD
1988-89	675.1	731.9	248004110.4	331867750.8	9032.8	8744.8	41.6	40.0
1989-90	664.8	775.8	196096787.7	278643809.9	10119.2	9411.0	39.6	42.5
1990-91	523.5	709.2	139839276.5	232499253.4	8754.8	9415.0	33.4	42.6
1991-92	485.8	683.2	115586233.0	200004279.4	6924.1	8982.2	34.6	47.6
1992-93	358.9	619.4	83039002.0	176293134.6	4514.3	7863.3	21.9	35.9
1993-94	339.6	626.7	67069753.7	153596169.6	3527.0	7080.0	20.0	36.4
1994-95	499.3	1225.5	63684821.1	136549938.9	11924.2	55541.0	20.7	36.3
1995-96	406.9	1041.3	60212162.1	146128944.1	11771.6	54625.9	17.9	35.4
1996-97	405.9	1096.5	55008691.0	136773913.3	10560.9	48512.0	19.9	42.0
1997-98	455.9	1257.3	54452733.8	134826712.7	11123.1	50778.7	21.2	45.4
1998-99	406.3	1024.1	45387523.9	114010314.8	13545.2	66903.7	20.3	45.7
1999-00	404.8	1106.4	57486770.5	151559690.6	11526.5	53638.8	17.0	37.1
2000-01	419.2	1219.7	66750916.5	166558985.8	11021.4	53041.7	16.1	34.1
2001-02	418.5	1135.6	54975952.1	139052967.0	11653.4	54743.6	16.0	34.4
2002-03	479.7	1446.3	56126217.8	142889616.6	13728.2	57804.4	16.1	34.0
2003-04	423.3	1196.2	58893556.1	148183067.3	14222.7	61122.9	15.7	33.8
2004-05	514.8	1529.6	64864335.0	161282808.3	18558.0	87090.6	16.5	35.6
2005-06	565.5	1586.7	59510121.1	145938097.6	21459.9	96858.0	16.6	34.5

Table-A3.4 Weighted Average TFP Levels for the aggregate mining industry and its four sub-sectors during 1988-89 to 2005-06

Year	Total Mining Industry	Metallic	Non-Metallic	Coal	Petroleum
1988-89	4.37	4.43	5.84	4.36	3.95
1989-90	4.90	5.35	5.03	4.57	6.62
1990-91	4.93	5.50	4.76	4.65	6.30
1991-92	5.10	5.88	5.54	4.69	6.34
1992-93	5.19	5.68	5.42	4.76	6.91
1993-94	5.51	5.64	6.33	5.07	6.83
1994-95	5.90	6.29	6.33	5.44	6.48
1995-96	5.47	6.13	6.15	5.15	5.30
1996-97	5.59	6.34	6.41	5.08	5.57
1997-98	5.66	6.07	6.51	5.30	5.55
1998-99	5.64	5.96	6.18	5.20	6.12
1999-00	5.50	6.25	6.34	5.08	5.16
2000-01	5.57	6.42	6.77	4.83	5.57
2001-02	5.85	6.68	6.69	5.16	6.13
2002-03	5.95	6.91	7.34	5.22	5.57
2003-04	5.81	6.22	7.78	5.06	5.32
2004-05	6.47	6.65	8.81	5.17	6.89
2005-06	6.55	6.35	9.09	5.50	6.73

CHAPTER FOUR

FIRM OWNERSHIP AND ENVIRONMENTAL COMPLIANCE

4.1 Introduction

This chapter compares the environmental performance of public- and private-sector mining firms. The study focuses on the Indian chromite mining industry and compares the environmental performances of public and private mining firms on four environmental indicators—management of overburden dumps, air quality in the locality, quality of drainage water after treatment, and quality of ground water.

The selection of the chromite mining industry as an example was prompted by the following reasons. Chromite mining generates an enormous quantity of hazardous pollutants such as hexavalent chromium Cr(VI) and other metals that contaminate air and water. So much so there was a fierce controversy when the Sukinda valley in Jajpur district of Odisha, the sole chromite-producing region in India, was rated as the world's fourth most polluted place by the Blacksmith Institute in the US in 2007. In response to the uproar in the media and civil society, the State Pollution Control Board (SPCB) of Odisha came out with a report denying the allegations of the Blacksmith Institute. The SPCB argued that the data source and methodology relied on by the Blacksmith institute were unreliable. It assessed the performance of chromite mining firms on four indicators—overburden management,[1] quality of mine drainage water, air pollution, and Cr(VI) in different water sources – separately.

[1] While extracting minerals, mining firms remove the top soil, or in the case of open-cast mining other unwanted soil from the pit, and dump it somewhere else. In chromite mining, this unwanted soil contains the hazardous hexavalent chromium. Run-off of this soil in rain water may pollute water bodies in the periphery and siltation may damage agricultural land. Therefore, before beginning operations, mining firms must submit a plan to the Indian Bureau of Mines on how they will manage the overburden dump and minimise environmental damage. This is usually

However, the performance of each of the 13 mining firms in the valley varied in different pollution indicators. Therefore, one is constrained to point out that it is hard to conclude that a firm fully complying in one indicator is doing the same in the case of all the other indicators. An assessment of the overall environmental performance of mining firms therefore needs a multidimensional measure, not just a unidimensional one.

While assessing the environmental performances of public and private firms, the existing body of literature tends to use just one of the environmental indicators (for example, air pollution or water pollution). Nonetheless, firms could generate pollution in more than one dimension. There is no doubt that the firms generating pollution in three dimensions are more harmful than the firms generating pollution only in one dimension. Similarly, the intensity of pollution might vary across firms. It would be misleading to assume that a firm polluting in one indicator will do so in other indicators, or a firm performing well in one indicator will do so in other indicators. Therefore, we need a new measure to assess the environmental performance of firms, one which can capture more than one pollution indicator (breadth) and reflect the intensity of pollution. This study presents a new methodology to assess environmental performance in a multidimensional framework. The new measure devised here is monotonic with respect to the dimension and intensity of pollution. Such a measure has great significance on the practical plane. For example, while imposing a penalty on a polluting firm it is crucial to measure the dimensional failures and intensity of pollution. Use of unidimensional measures will impose either a nil or full penalty on the firm. The application of the new methodology is here demonstrated in the context of the Indian chromite mining industry.

The new measure, called the Multidimensional Environmental Defiance Index (MEDI),[2] to assess the environmental compliance of mining firms uses four pollution indicators—(i) quality of mine drainage water,[3] (ii) overburden management, (iii) suspended particulate matter

---

achieved by stabilising the dumps with plants, fixing mats or constructing retention walls.

[2] For a multidimensional index in poverty literature, see Alkire and Foster (2008).

[3] Open-cast chromite mining generates a huge volume of seepage water containing hazardous Cr(VI) . If mine drainage water containing Cr(VI) is released untreated, it can seriously contaminate nearby water bodies. Many mines have chrome ore beneficiation (COB) plants, where the chromium content in the ore is concentrated through washing and sorting. COB plants can also be a source of Cr(VI). Mining

(SPM) in ambient air,[4] and (iv) Cr(VI) in drinking water.[5] Although our analysis is based on only 10 mining firms, they produce around 90% of India's chromite.

The rest of this chapter is organised as follows. Section 4.2 reviews the theoretical and empirical literature that analyses the relationship between firm ownership and environmental performance. Section 4.3 explains the methodology and data sets used for the present analysis. The results of the analysis are presented in Section 4.4 and Section 4.5 concludes with a summary of the findings.

4.2. Ownership and environmental performance

Do private and public firms comply with environmental regulations differently? Both the theoretical and empirical literature are highly divided on this. In the following section, we present a summary of the selected literature examining the relationship between firm ownership and environmental performance.

4.2.1 Theoretical literature

Each firm attempts to minimise its total cost subject to an output target (constraint). The total cost of a firm has three components—(i) input cost, including the expenditure on pollution abatement; (ii) penalties paid to the regulatory authorities; and (iii) the social cost accrued due to pollution. Private firms, however, consider only the first two cost components and treat the third component as zero. But public firms are presumed to account for the social cost. The decision of a firm to abate environmental damage is analysed with a simple optimisation model in Wang and Jin (2002). Keeping in view these three costs of production, the difference in

---

firms are directed to release this water only after treatment to contain any hazardous pollutants.

[4] Open-cast mining operations and the transportation of minerals by heavy vehicles causes serious air pollution in mining regions, mineral dust being the major source. Mining firms are instructed to control this by spraying water on roads and adopting better blasting and processing technology. The air pollution is measured in terms of micrograms of SPM per cubic metre of air. The permissible limit set by the Odisha SPCB is 500 micrograms of SPM per cubic metre of air in industrial areas and 200 micrograms in residential areas.

[5] Seepage of mining water into the ground water table might cause serious problems when people drink this water. The quality of ground water is measured in terms of milligrams of Cr(VI) per litre and the permissible limit is 0.050.

environmental performance of public and private firms would emanate from three channels.

i) Regulation Effect: If only government's environmental regulations are considered, the environmental performance of private firms would be better than that of public firms. The primary reason for this would be the better bargaining power of the public sector firms with the government authorities. However, rent-seeking activities by private firms would reverse the scenario. Managers of public firms are left with little scope to bribe the regulatory agencies to wriggle free of punitive actions. Managers of private firms, however, could bribe the regulatory agencies to escape from the oversight of regulatory agencies. In such a context, private firms would be more polluting than their public counterparts.

ii) Internalisation Effect: Assume the strength of environmental regulation is the same for all companies, and the only difference in the marginal prices of inputs is caused by the internalisation of the pollution externality. Private firms do not pay heed to the environmental cost and would use more pollution-generating inputs compared to state-owned firms, which are conscious of their social responsibility. Therefore, public firms would perform better than private firms.

iii) Efficiency Effect: Production efficiency could cause differences in environmental performance. For an input positively contributing to pollution discharge, a higher efficiency means a lower marginal discharge of that input. For an input helping to abate pollution, higher efficiency means higher marginal pollution reduction. Therefore, higher efficiency means less pollution generation and high pollution reduction, and finally better environmental performance. Depending on productive efficiency, the environmental performances of public and private firms could vary.

The overall environmental performance of a firm will, therefore, be the sum of the three effects discussed above.

Friedman (1970) argues that the sole objective of business is to maximise profit.[6] Consequently, private firms are believed to be bad environmental performers compared to public firms whose basic objective is to maximise social benefits. In contrast, it is asserted that publicly owned plants are quite likely to be older, less efficient and therefore more pollution-intensive than their private counterparts. We might expect a lower pollution intensity in the case of public plants operating under soft budget constraints because they are in a position to spend more to check

[6] "There is one and only one social responsibility of business—to use its resources and engage in activities designed to increase its profits so long as it stays within the rules of the game, which is to say engages in open and free competition without deception or fraud" (Friedman, 1970).

pollution. However, bureaucratic control may shield state-owned facilities from local pressure. The empirical findings of Pargal and Wheeler (1996) reveal that public ownership is strongly associated with dirty production and hence the effect of bureaucratic shielding seems to outweigh any leverage from soft budgets. The differential environmental performances of public and private firms could accrue through three channels—(i) efficiency; (ii) size of the firm; and (iii) regulatory differences.

A number of studies have attempted to explain the environmental effects of production inefficiencies (Pearce et. al., 1990; O'Connor, 1991; Warhurst, 1999; Loayza, 1999). They point out that low investment capacity constrains firms to accumulate non-resource capital, develop organisational capabilities and skilled human resources. Increased competitiveness encourages investment in technological capability and production capacity, and this in turn reduces pollution per unit of output, whereas decreased competitiveness increases pollution per unit of output. The most efficient firms are generally better environmental managers because they are innovators. Furthermore, where the costs of complying with environmental regulations threaten competitiveness, dynamic firms can offset these costs by improving production efficiency. O'Connor (1991) points out that in the mining industry, low educational and skill levels of workers can negatively affect productivity and the maintenance of equipment. This reduces profit and constrains a company's capacity to invest. As a result, companies are unable to renew capital equipment or acquire state-of-the-art equipment that pollutes less per unit of output.

Hilson (2000) points out that economic and technological barriers may cause differences in environmental performance. Even though there are highly efficient waste minimisation technologies available in the market, lack of funds prevents widespread adoption of these in the mining industry. In many instances, structural barriers prevent the adoption of cleaner technologies and strategies. Some of the pollution-control systems at sites represent billion-dollar investments, and the employees have skills and knowledge specific to the system. Changes to conventional technologies could make workers and managers obsolete, and would require investment by companies in training programmes, an added difficulty for a firm with a limited budget. In many parts of the world, another major technological barrier preventing the adoption of cleaner technology in mining operations is the lack of available systems.

Although private firms may have higher efficiency in resource utilisation, they may not seek to internalise environmental costs (Baumol and Oates, 1988). In other words, the private sector may compromise the environment to avoid the potential cost of environmental investments and expenditures. However, public firms would seek to internalise environmental costs because their objective is to maximise social welfare. Environmental performance could also vary a great deal owing to the different levels of "environmental bargaining power" of firms. Wang and Jin, 2002, define environmental bargaining power as "an enterpriser's capacity to negotiate with the local or national environmental agencies pertaining to the enforcement of pollution control regulations such as pollution charges, fines, etc". The managers of public firms and private firms will have different bargaining strengths with the pollution control authorities. This would result in differences in environmental performance (Wang and Jin, 2002). Also, firms with different ownership may receive different levels of informal regulations, or community pressure, on pollution abatement.

4.2.2 Empirical literature

McMahon et. al. (1999) in a study of artisanal, small and medium mining firms in Bolivia, Chile and Peru point out that though averages show the artisanal and small mine sector is significantly dirtier per unit of output than other types of mining, this is not always the case. Comparing the environmental performances of medium and large-scale mines, the study does not find any clear distinction between them to be as important as the distinction between old and new mines. In Chile, though the state-owned Empresa Nacional de Mineria (ENAMI) is entrusted with smelting all the output of small and medium producers, its smelters are heavy polluters and there have been a number of law suits against it. This implies that ownership does not make any significant difference to environmental compliance. MacMahon et. al. (2000) in a study of mining in Indonesia demonstrate that the environmental performance of medium-scale mines is considered poor to very poor. Artisanal and small-scale mining in Indonesia is also undertaken with little or no environmental care. But large-scale mines in the country seemed to be using state-of-the-art technologies and practices and had a relatively limited impact on the environment.

Chakravorty (2001) observes that small-scale mines in India, particularly the very small ones, normally do not bother about eco-friendly operations. They not only inadvertently destroy (also deliberately at times for extra income) vegetation and trees, particularly at and near their area of operation, but also do not take any steps to compensate for it. Ghose (2003) points out that approximately 90% of India's mines are operating on a small scale. Of them, only a few are semi-mechanised and the rest are predominantly manual. Improper exploration techniques, lack of planning and low to intermediate technology result in the poor recovery of mineable deposits. Environmental protection in these mines is seldom more than rudimentary.

The effect of community pressure on emissions has been confirmed in several empirical studies (Pargal and Wheeler, 1996; Wang, 2000). They find that proxies for direct community pressure (community income and education levels) have significant effects on plant-level emissions. But whether a community takes environmental action or at what level informal regulation and community pressure are effective possibly depends on the impact a certain enterprise has on the regional economy. There is an inherent trade-off involved in local residents choosing an optimal pressure level to apply on a certain enterprise because they take into consideration the potential economic benefits from job opportunities and income expectations. Nunez-Barriga and Castaneda-Hurtado (1999), in the context of Peru, say that environmental behaviour is unrelated to ownership structure (foreign, state, or domestic private) or the size (large or medium) of firms. However, the longevity of production capacities has significant bearing on environmental compliance.

The review of theoretical and empirical literature reveals that while assessing the environmental performances of public and private firms, the existing body of literature tends to use just one of the environmental indicators (for example, air pollution or water pollution). Nonetheless, firms could generate pollution in more than one dimension. There is no doubt that the firms generating pollution in three dimensions are more harmful than the firms generating pollution only in one dimension. Similarly, the intensity of pollution might vary across firms. It would be misleading to assume that a firm polluting in one indicator will do so in other indicators, or a firm performing well in one indicator will do so in other indicators. Therefore, we need a new measure to assess the environmental performance of firms, one which can capture more than one pollution indicator (breadth) and reflect the intensity of pollution. Moreover, the review of theoretical and empirical literature on firm ownership and environmental performance does not provide us with any single conclusion and this issue remains as an empirical issue.

4.3. Methodology and Data

4.3.1 Methodology

To compare the environmental performances of public and private firms, the study focuses on the Indian chromite mining industry.[7] First, a comparison is made, separately, on four indicators: i) quality of mine drainage water, ii) management of overburden, iii) ambient air quality and iv) quality of drinking water. In the second step, an aggregate environmental performance measure named the Multidimensional Environmental Defiance Index (MEDI) has been constructed and a comparison is made between the environmental performances of public and private firms through a permutation test. Before venturing into the comparison, we explain the MEDI in detail and its computation process.

In a very simplistic framework, a multidimensional environmental performance can be measured by aggregating all the indicators. As the indicators will be in different units, we will have to normalise them by converting them into indices. Next we can compute the average of all the indices and arrive at a single indicator for the overall environmental performance of a firm. However, this simple averaging of indices suffers from several limitations. From a regulator's perspective, it is crucial to examine whether a firm has complied with an environmental standard or not. If not, to what extent has it exceeded the permissible limit? To target polluting firms, it is therefore more important to identify the defiant firms and measure their degree of defiance (intensity of pollution) than measure the degree of compliance of obedient firms. This is called the focus axiom. Thus we focus only on the defiant firms and gauge the intensity of their defiance. Simple indices (those like the popular Human Development Index), however, fail to perform in this manner. In the process of simple averaging, an extra level of achievement in compliance automatically offsets an extra level of defiance.

Although the merits of an extra level of achievement in compliance cannot be undermined, it would be unwise to use it to make up for an extra level of defiance, at least from a regulator's point of view. The measurement of an extra level of compliance like keeping pollution far below the permissible limit might be worthwhile in the context of carbon trading. However, here our focus is from a regulator's standpoint. From this, a firm's extra level of achievement within permissible limits does not

---

[7] Chromite mining is one of the most polluting industries. For more on the environmental effects of chromite mining, see Appendix A4.1.

make much of a difference. But if a firm exceeds the permissible limit, it causes serious environmental and health damages. With an increase in the level of defiance, the level of damage also goes up. Moreover, an extra level of compliance in one indicator cannot compensate for violation in other indicators. For example, a firm might have kept the level of SPM in the air 10 points below the permissible limit but exceeded the level controlling Cr(VI) in drinking water by 10 points. Here, it would be inappropriate to say that the 10-point extra achievement in controlling air pollution offsets the 10-point extra defiance in failing to control the level of Cr(VI) in drinking water. Therefore, while measuring the environmental performances of pollution-generating firms, they should first be classified into compliant and defiant. After this, the degree of defiance of defiant firms should be gauged. In the following section, we explain the methodology of constructing the MEDI in detail.

The measurement of multidimensional environmental defiance is carried out in two stages. The first stage involves identification of defiant firms, which is done by setting the criteria for distinguishing defiant firms from obedient ones. The second stage involves assessing the dimensional failure and measuring the intensity of pollution. In the identification exercise, we take the permissible limit set by the SPCB as the first cut-off point, following which we categorise all firms into obedient or compliant (firms keeping emissions below the permissible limit) and defiant (firms exceeding the permissible limit for emissions) in each pollution indicator. However, to arrive at a single indicator, or conclude whether a firm is on the whole defiant or compliant, we have to set a dimensional cut-off point. For this, we count the number of indicators a firm has failed to comply with. Then depending on the dimensional cut-off, we can categorise them into compliant and defiant. In a strict regulatory environment, a firm will be considered to be defiant if it exceeds the permissible limit in at least one indicator, and in a full liberal regime, if it violates the limits in all the indicators. However, we can set the cut-off point in between these two polar extremes. After identifying the defiant firms, we measure their degree of violation by measuring the intensity of pollution.

Step-1: Identification of defiant firms with the help of permissible limits

Let a_{ij} be the environmental performance of firm i in j^{th} indicator/dimension. $d \geq 2$ be the number of environmental dimensions under consideration. Let l_j denote the permissible limit of jth indicator and $l(l_j \in l)$ is a row vector of $1 \times d$ dimension. Firm i will be considered to be defiant in dimension j if $a_{ij} > l_j$. Following the permissible limit we can identify the compliant and defiant firms in each indicator separately and let us mark 1 for the firms who have exceeded the permissible limit and 0 for the firms who have kept the pollution below the permissible limit. If there are *N* number of firms, we arrive at a $N \times d$ dimensional matrix of 1 (defiant) and 0 (compliant firms). Here ends the first step of identification on the basis of permissible limit.

Step-2: Identification of defiant firms with the help of dimensional cut-off

The second step of identification involves labelling the firm defiant or compliant on the basis of dimensional cut-off. As mentioned earlier, we can set the dimensional cut-off C_k (where, $1 \leq k \geq d$) depending upon the regulatory regime. Depending upon the dimensional cut-off, a firm will be considered to be defiant if it exceeded the permissible limit in more than the *k* dimensions. Thus, we arrive at a $N \times 1$ dimensional column vector (I) with 1 (defiant) and 0 (compliant firms) parameters.

Step-3: Measuring the breadth of defiance

The third step involves measuring the breadth of the defiance. This is simply done by counting each firm's defiance and dividing it by the total number of dimensions/environmental indicators (d) Let *k* be firm *i*'s total defiance count out of the total *d* number of dimensions. The ratio of firms *i*'s defiance (S_i) can be computed as *k/d*. The value of S_i ranges between 0 and 1. This generates a $N \times 1$ dimensional column vector.

Step-4: Measuring the intensity of defiance

Fourth step of the computation involves measuring the intensity of defiance, E_{ij}, (extent to which the firm has exceeded the permissible limit). If the environmental indicators represented in cardinal form, we can easily measure the intensity of defiance. This exercise is carried out only for the defiant firms (recall here the focus axiom!). The values of intensity of defiance are derived by deducting the permissible limit values from the actual values. We can write this in notation form as $E_i = a_{ij} - l_j$ if and only if $a_{ij} > l_j$. In order to make the intensity of defiance values scale free we can normalise these values using the simple formula $\frac{actual - \min imum}{\max imum - \min imum}$. This generates a $N \times d$ dimensional matrix of decimal values (normalised intensity of defiance) and 0 (compliant firms). By assigning equal weights (or differential weightage depending upon the harmful effect of each indicator) to all environmental indicators we can arrive at a single value for each firm's intensity of defiance by taking the simple average of the normalised E_{ij}. In notational form we can write this as $G_i = \frac{\sum E_{ij}}{d}$. Thus, we arrive at a $N \times 1$ dimensional column vector of average intensity of defiance.

Step-5: Computing Multidimensional Environmental Defiance Index (MEDI)

The fifth and last step involves the computation of Multidimensional Environmental Defiance Index (MEDI) using all the information obtained from step 3, 4 and 5. Thus, *MEDI* = *I* x *S* x *G*. The values of *I* reflects whether the firm is identified as a defiant or compliant firm on the basis of dual cut-off (first cut-off is permissible limit and second cut-off is environmental dimensions) approach, values of *S* represents the breadth of defiance and *G* reveals the intensity of violation. The values of *MEDI* range between 0 and 1 and we arrive at a *N* x 1 dimensional column vector. The values of *MEDI* increases with the increase in the values of *G* and *S*. Therefore, it satisfies the property of monotonicity. Further the index satisfies the focus axiom. In other words, *MEDI* represents the values only for defiant firms which are identified by dual cutoffs.

4.3.1.1 Ordinal and Cardinal Data

The merit of this multidimensional environmental defiance index is that it allows us to use both cardinal and ordinal data together. The mixed case poses no problems for the dual cutoff identification method (C_k) nor for *S*. However, for *G*, a tension arises across dimensions: they cannot be applied to ordinal dimensions and yet dichotomisation of cardinal dimensions loses valuable information. In such situations, there may be grounds for creating a hybrid defiance matrix in which entries are normalised excesses for the cardinal dimensions and 0-1 defiances for the rest. The monotonic measure *G* can then be computed from this matrix to obtain measures that reflect the intensity of defiance in each cardinal dimensions, but follow the ordinal measurement restriction for the remaining dimension. Even in case of ordinal variables with more than two rankings we can still compute the normalised excess where values lie between 0 and 1and we can incorporate this into the computation exercise for *G*. This process may seem to increase the effective weight on ordinal dimensions since all defiant firms will appear to have the most severe degree of defiance possible. As a correction, differential weights across dimensions could be assigned.

4.3.2 Data

To examine the relationship between firm ownership and environmental performance, the study focuses on chromite mines in Odisha, a state in the east of India. Odisha has about 98% of the total proven chromite (chromium ore) reserves in the country, of which about 97% is in the Sukinda valley in Jajpur district, stretching over an area covering approximately 200 square kilometres (CSE, 2008). There are 13 chromite mining firms operating in the Sukinda valley, of which systematic data are available for 10. These 10 mining firms contribute around 90% of the total chromite production in the valley. Thus our data set represents around 88% of India's total chromite production. Data pertaining to the annual production of chromite, year of commencement of mines, and environmental compliance on four indicators have been gathered from the Odisha SPCB in Bhubaneswar. The four environmental indicators on which data have been collected are: (i) the quality of mine drainage water (collected between 2005 and 2007 at various points of time for different firms), which shows the extent of Cr(VI) (milligram per litre) in the water draining out from the effluent treatment plants; (ii) the management of overburden by mining firms; (iii) the ambient air quality in industrial areas (collected between April 2004 and April 2005), which measures the level of SPM in micrograms/cubic metre; and (iv) the quality of drinking water, which shows the level of Cr(VI) in drinking water in nearby borewells between April 2004 and April 2005. The management of overburden by mining firms is classified into three categories—satisfactory, partial and poor—with ranks of 1, 2 and 3. For the other three indicators, the maximum values of pollutive content have been taken from the sample.

4.4. Results

In this section, we present the results in two parts. In the first part, the comparison between the environmental performances of public and private firms has been made with the help of single indicators. In the second part, the comparison is made with the help of the MEDI.

4.4.1 Unidimensional Environmental Performance

The data pertaining to the four pollution indicators of 10 chromite mining firms are presented in Table 4.1. Firms' pollution level in three indicators—quality of drainage water, air and drinking water—has been compared with the permissible limit to categorise them as compliant or defiant. Except for the management of overburden, the data for the other three indicators are cardinal in nature. The data for the management of overburden are ordinal in nature, and the performance of firms has been categorised as satisfactory, partial, and poor with ranks of 1, 2 and 3 respectively. Firms with a satisfactory performance have been categorised as compliant (0) and rest as defiant (1) (see Table 4.2). The last column in Table 4.2 shows the number of indicators in which each firm has transgressed limits. The comparison of public and private firms in each indicator does not show any significant difference between the two groups. A comparison of defiance counts also does not demonstrate any clear distinction between the two groups.

Comparing each of the four indicators separately does not allow us to conclude whether a firm is defiant or compliant. The simple classification of firms into compliant and defiant also does not reflect the breadth and intensity of defiance. The MEDI overcomes these shortcomings and enables us to compare the aggregate environmental performance of each firm.

Table 4.1 Environmental Performance of Chromite Mining Firms in Odisha

Firm Code	Owner	Drainage Water	OBM	Air Quality	Drinking Water
1	Private	0.22	2	289	0.006
2	Private	1.025	3	171	0.022
3	Public	0.69	2	247	0.025
4	Private	0.444	3	260	0.004
5	Private	1.02	2	280	0.033
6	Private	0.119	1	272	0.035
7	Public	0.2	3	316	0.001
8	Public	0.108	1	208	0.004
9	Public	0.444	2	161	0.028
10	Private	0.169	1	600	0.04
Permissible Limit		0.100		500	0.05

Source:
Odisha State Pollution Control Board (2007), Bhubaneswar.
Notes:
Quality of drainage water is measured in micrograms of Cr(VI) in a litre of water flowing out after treatment.
Overburden management (OBM) has been categorised into 1=Satisfactory, 2=Partial, 3= Poor
Air quality is the level of SPM in micrograms per cubic metre of air.
Quality of drinking water is measured in micrograms of Cr(VI) in a litre of water.
Values exceeding the permissible limit are considered to be harmful.

Table 4.2 Identification of Defiant Mining Firms

Firm Code	Owner	Drainage Water	OBM	Air Quality	Drinking Water	Defiance Counts
1	Private	1	1	0	0	2
2	Private	1	1	0	0	2
3	Public	1	1	0	0	2
4	Private	1	1	0	0	2
5	Private	1	1	0	0	2
6	Private	1	0	0	0	1
7	Public	1	1	0	0	2
8	Public	1	0	0	0	1
9	Public	1	1	0	0	2
10	Private	1	0	1	0	2

Source: Own computation
Note:1 = Defiant, 0= Compliant

In the second part, we present the values of the MEDI, which is monotonic with respect to dimension (breadth) and intensity, and compare the aggregate environmental performances of public and private firms.

Table 4.3 Identification of Defiant Firms on Dual-Cut-off Criterion

Firm Code	Owner	I1	I2	I3	S_i
1	Private	1	1	0	0.50
2	Private	1	1	0	0.50
3	Public	1	1	0	0.50
4	Private	1	1	0	0.50
5	Private	1	1	0	0.50
6	Private	1	0	0	0.25
7	Public	1	1	0	0.50
8	Public	1	0	0	0.25
9	Public	1	1	0	0.50
10	Private	1	1	0	0.50

Source: Own computation
Note: 1 = Defiant, 0= Compliant

4.4.2 Multidimensional Environmental Performance

At the outset, all mining firms have been categorised as compliant and defiant by following the dual cut-off approach. In Table 4.3, three indices, I_1, I_2 and I_3, categorise the firms as defiant if they exceed the permissible limit in one, two or three indicators. The last column (S_i) represents the breadth of defiance or dimensional failure of each firm, which is computed by taking the ratio of the firm's defiance count to the total number of dimensions. A comparison of the performances of public and private firms in these indicators does not show any significant difference. However, until now, we have only categorised the firms as complaint and defiant without measuring their degree of defiance. This has been addressed in the MEDI. The values of the MEDI are presented in Table 4.4.

As mentioned earlier, the MEDI satisfies the monotonicity property with respect to breadth and intensity of defiance. A firm flouting more indicators and doing so far beyond the permissible limits will score higher values in the index. The $MEDI_1$, $MEDI_2$ and $MEDI_3$ present the indices values with a one, two and three dimensional cut-off respectively. The differences between the three measures of the MEDI on the basis of different cut-offs is clearly seen in Table 4.4. With a rise in the number of dimensions in the dual cut-off approach, the number of defiant firms declines. This is seen when we compare the $MEDI_1$ and $MEDI_2$. According to the $MEDI_1$, all firms are defiant, whereas only eight firms are defiant according to the $MEDI_2$. More interestingly, not a single firm is defiant according to the $MEDI_3$.

A comparison of the MEDI values of the environmental performances of private and public firms also does not show any significant difference. Of the 10 chromite mining firms, six are private and four are public. As per $MEDI_1$, all public and private firms are defiant and the mean MEDI value for private firms is 0.136 and public firms is 0.097.

To test the statistical significance of the difference between the mean MEDI values of public and private mining firms, we use a permutation test. Permutation test[8] allows us to examine the statistical significance of the differences between the two groups with small samples without making any distributional assumption. It applies computing power to relax some of the conditions needed for a traditional inference and to infer in new settings.

[8] See Hesterberg et al (2003) for an elaboration on the permutation test.

Table 4.4: Multidimensional Environmental Defiance Index

Firm Code	Owner	$MEDI_1$	$MEDI_2$	$MEDI_3$
1	Private	0.078	0.078	0.0
2	Private	0.250	0.250	0.0
3	Public	0.142	0.142	0.0
4	Private	0.171	0.171	0.0
5	Private	0.187	0.187	0.0
6	Private	0.001	0.000	0.0
7	Public	0.138	0.138	0.0
8	Public	0.000	0.000	0.0
9	Public	0.108	0.108	0.0
10	Private	0.133	0.133	0.0

Source: Own computation

The test proceeds as follows. First, the difference in means between the two samples is calculated—this is the observed value of the test statistic, T(obs). Then the observations of groups A and B are pooled. Next, the difference in sample means is calculated and recorded for every possible way of dividing these pooled values into two groups of size n_A and n_B (that is, for every permutation of the group labelled A and B). The set of these calculated differences is the exact distribution of possible differences under the null hypothesis that the group label does not matter. The one-sided p-value of the test is calculated as a proportion of sampled permutations where the difference in means is greater than or equal to T(obs). The two-sided p-value of the test is calculated as a proportion of sampled permutations where the absolute difference is greater than or equal to T(obs).

Table 4.5: Summary of the MEDI Values for Public and Private Chromite Mining Firms

	Mean MEDI	N	SD
Public Firms	0.097	4	0.067
Private firms	0.136	6	0.088

Monte Carlo Permutation Results
Replications 100

	T(obs)	C	N	P=C/N	SE
Sum	0.819	29	100	0.290	0.045

Note: SD- Standard Deviation
Confidence interval is with respect to P=C/N.
$C = \#\{|T| >= |T(obs)|\}$

In the present study, a permutation test with 100 repetitions shows that 29 of the 100 randomly permuted data sets yielded sums from the private group larger than or equal to the observed sum of 0.819 (see Table 4.5). This suggests that at 5% level we cannot reject the null hypothesis that there is no difference between the mean MEDI values for public and private mining firms. This implies that there is no significant difference between the environmental performance of public and private mining firms.

4.5. Conclusion

In this chapter, we examined the highly debated question of whether public and private firms comply with environmental regulations differently by undertaking an empirical exercise in the context of the Indian chromite mining industry. It proposed a new methodology to measure environmental performance in a multidimensional framework. The merit of a multidimensional measure over a unidimensional one for comparing the environmental performance of pollution-generating firms is clearly revealed in the analysis. Although our analysis is based on a few mining firms, they produce around 90% of India's chromite. Moreover, to overcome the small data set constraint, a permutation test has been carried out to ascertain the statistical difference. A comparison of the environmental performances of public and private mining firms in four indicators separately gives inconclusive results. However, the MEDI, which satisfies the property of monotonicity with respect to the breadth and intensity of violations, facilitates a better comparison. With the help of both unidimensional and multidimensional indices, the study does not find any significant difference in the environmental performances of public and private mining firms.

Appendix

A4.1. Environmental Impacts of Chromite Mining

Chromium is a metallic element with an atomic number of 24. It is a member of sub-group VIa on the periodic table, along with molybdenum, uranium and tungsten. Chromium generally occurs in small quantities associated with other metals, particularly iron. The most common prevalences are +3 and +6. Chromium forms a number of salts, which are characterised by a variety of colours, solubilities and other properties. The name "chromium" is from the Greek word for colour. The most important chromium salts are sodium and potassium chromates and dichromates, and potassium and ammonium chrome alums. The metal is usually produced by reducing the chromite (FeCr2O4) ore with aluminium. Chromium is used to harden steel to manufacture stainless steel, and in the production of a number of industrially important alloys (Weast et. al., 1988). It is used to make pigments, in leather tanning and for welding. Chromium plating produces a hard mirror-like surface on metals that resists corrosion and adds to appearance.

Cr(III), as found in chromite and other naturally occurring minerals, is an essential micronutrient for normal glucose metabolism. Chromium deficiency can lead to insulin circulation and cardiovascular problems. There are reports that even relatively large doses of Cr(III) do not have any harmful effect when fed in water or food to animals. What is not absorbed in the gastrointestinal tract is excreted. It is believed that Cr(VI) is formed by human activities. It is rapidly reduced to relatively harmless Cr(III) in acidic solutions ($pH < 4$) by organic matters or biomass. Beyond a certain concentration, Cr(VI) is toxic, inducing symptoms such as skin ulcers, vomiting, diarrhoea, gastrointestinal bleeding, and maybe even cardiovascular shock. It is cytotoxic, mutagenic and carcinogenic. For long, the chromite matrix was considered to be quite stable in the Cr(III) state. However, recent studies reveal that Cr(III) lodged in chromite can oxidise to toxic Cr(VI) though various physio-chemical and biological processes. Chromite occurs either in a lumpy or friable form. It has been observed that the Cr(VI) problem is generally associated with mining the friable mineral.

A.4.1.1 Presence of Cr(VI) in mine drainage water

Open-cast chromite mining generates a huge volume of seepage water. Even though chromium in chromite is in a trivalent state, some hexavalent Cr(VI) is always formed due to certain complex reactions. If mine drainage water containing Cr(VI) is released untreated, it can severely contaminate nearby water bodies. Many mines have chrome ore beneficiation (COB) plants, where the chromium content in the ore is concentrated through washing and sorting. So COB plants can also be a source of Cr(VI).

A.4.1.2 Overburden generation

Open-cast chromite mining results in enormous quantities of overburden. The stripping ratio varies from 1:5 to 1:10. Unless managed properly, run offs from the overburden dumps have the dual potential of polluting water bodies by siltation and leaching of Cr(VI).

CHAPTER FIVE

FIRM OWNERSHIP AND SOCIAL COMPLIANCE

5.1. Introduction

Do public and private mining firms acquire private land and compensate tenants in a similar fashion? Or are there some differences? A cursory look at the ongoing land acquisition process in India gives the impression that private firms face greater opposition when trying to acquire private land. The strong protests against new projects initiated by private firms like Pohang Steel Company (POSCO), Tata and Vedanta in Odisha, and Tata's projects in Singur in West Bengal are examples.[1]

Public protests against private mining firms could cause two kinds of problems. First, inadequate compensation to private landowners will accentuate the distributional problem and augment social chaos. Second, if the reasons for public protests against the private sector are not genuine, the economy will be placed in a Pareto inefficient position because of the disinterest of private firms to invest in the mining sector, thus reducing output.

Involuntary land acquisition for mining projects needs special attention because of the nature of mineral deposits. Minerals, otherwise known as point resources, are available only at specific locations. Therefore there are limited options to change the location of a mine from one place to another.

Empirical studies comparing the compensations provided by public and private firms are few and far between. Based on the theoretical literature from public economics, we can only make a few propositions. The inferences drawn from these propositions, however, present an ambiguous picture of the absolute merit of any sector. Therefore, the issue remains purely an empirical one. Given this, the present study seeks to compare the differences between public and private firms in providing compensation to those affected by land acquisition in the context of the Indian mining industry.

---

[1] For example, see Bhattacharya (2007); Chandra (2008); Fernandes (2007); Sarkar (2007).

Based on a survey of 69 households (making up 84 land transfer cases) that surrendered their land to mining firms in two mineral-rich districts (Kendujhar and Jajpur) of Odisha, the study compares the social compliance of public and private mining firms. The rest of this chapter is organised as follows. Section 5.2 defines the concept of social compliance and distinguishes it from Corporate Social Responsibility (CSR). Section 5.3 reviews the related literature. Section 5.4 outlines the theoretical propositions for the differential performances of public and private firms in social compliance. The methodology and data used for the analysis are discussed in Section 5.5. Section 5.6 presents the findings of the study. The findings of the study are discussed in section 5.7 and Section 5.8 has the concluding remarks.

5.2. Social Compliance: The Concept

The term "social compliance" is seldom found in economic literature. However, a large body of work related to, though not synonymous with, social compliance can be found in the literature on CSR. At the outset, it is imperative to define the concept of social compliance and distinguish it from CSR. The term social compliance has been defined as the compensation made to the individual households and/or communities as a whole for the direct and indirect economic, social and environmental losses accruing due to the involuntary land transfer. Here, the first question that arises is why involuntary land acquisition should be considered a social issue. Land acquisition would not be labelled a "social issue" if both parties—tenants and mining firms—are able to clinch a deal without any disagreements. However, such exchanges seldom take place. The government then invokes the "power of eminent domain" under the Land Acquisition Act (LAA) and acquires the desired land in the name of "public interest". Given the geographical specificity of minerals, it is sometimes legitimate for the government to invoke the LAA. Government intervention in land acquisition becomes essential when there is a mismatch between the reservation price of a tenant, which is an aggregation of the economic and non-economic values of land, and the amount a mining firm is willing to pay. In such cases, the government mediates between the tenant and the lessee, adhering to specific public policy guidelines, to see the project through.

The second question that arises is what norms a mining firm should comply with. Government guidelines on land acquisition have been a very contentious issue in India and other developing countries. Policies on land acquisition and compensation have changed over the years. Although the

Central government recently formulated a new policy on land acquisition, rehabilitation and resettlement, no law has been put in place. The concept of social compliance becomes vague in a context where both the "rules of the game" and the regulatory mechanisms are not in place.

Our goal is not to gauge the social compliance of public and private mining firms against some idealistic benchmark and work out the shortfall, but to make a comparison between the performances of public and private mining firms. Therefore, based on compensation received by the households that surrendered their land, and their satisfaction the study attempts to draw conclusions on whether public or private firms have done better in providing compensation.

5.2.1 CSR versus Social Compliance

Before getting into a comparison of social compliance by public and private mining firms, let us look into how the concept social compliance varies from the more popular concept of CSR.

Although CSR has become a buzz word among corporate professionals, it lacks a systematic definition. A widely accepted definition of CSR is provided by the World Business Council for Sustainable Development, 1998, which says it is "the continuing commitment by a business to behave ethically and contribute to economic development while improving the quality of life of the workforce and their families as well as local community and society at large". The European Commission has defined CSR as a "concept whereby companies integrate social and environmental concerns in their business operations and in their interaction with their stakeholders on a voluntary basis".

The Federation of Indian Mineral Industry (FIMI) in its policy statement defines CSR as an investment that provides a social licence to continue to operate a business, which is increasingly becoming very challenging. It points out that CSR results in a "win-win proposition for companies as well as for all its stakeholders, which ensures the sustainability in long term" (FIMI, 2009).

Drawing on Friedman (1970), one could argue that firms should focus on maximising their profit and pay legitimate taxes, which will ultimately be spent on a nation's social and economic overheads. As pointed out earlier, these arguments have not got enough support and private firms directly spend a sizeable sum on CSR activities. This is done on a voluntary basis and is often seen as a way of enhancing the reputation of a brand or company. The goodwill generated by CSR measures results in substantial positive returns—in terms of easy clearance of projects, more public faith in the shares of a firm and so on.

By social compliance we mean the direct compensation made to and facilities made for an individual or community for the direct and indirect economic, social and environmental losses caused due to the involuntary transfer of property, displacement and damage to the local natural resource base. Unlike CSR, social compliance is obligatory in nature. Social compliance, which is here synonymous with the compensation over involuntary land acquisition or displacement, is an ex ante measure, whereas CSR is an ex post measure. The nature of compensation is determined before the land is acquired. A part of the compensation—for example, in the form of cash, substitute houses or land— is paid before a project begins operating. Another part of the compensation—like employment in the project, or the expenditure of a fixed proportion of profit on local area development—is delivered after the project begins operations. If, during negotiations, a firm promises to spend a part of its net profit on local area development, a measure that would have been optional under CSR becomes obligatory. Social compliance thus becomes a broader concept than merely compensation.

5.3. Insights from Literature

How should tenants be compensated for the involuntary transfer of land to the mining firms? Does firm ownership differentiate the compensation package? Economic analysis of these questions is sparse. The structure of property rights and the nature of compensation vary across countries. Therefore uniform application of single theory will be inappropriate for different countries.

Mining policy and legislation for fuel minerals differs from that for non-fuel minerals. Moreover, the fuel-mineral sector in India is dominated by public-sector mining firms. The present study focuses mainly on non-fuel minerals. Any mining firm (private or public) in India needs to obtain two different rights—the right to extract a mineral, which is sanctioned by the state or central government;[2] and the right to the surface of the land

[2] Minerals in India are considered to be national property. The Mines and Minerals (Regulation and Development) Act, 1957 empowers the Central government to make necessary legislation and regulate the production, conservation and development of the major minerals defined in Schedule A. Respective state governments are provided with the power to regulate, conserve and develop all the minor minerals listed in Schedule B. Land being in the state list, the onus of issuing surface rights is vested in state governments.

under which the mineral lies. Depending on the ownership of a particular plot of land, mining firms can obtain the surface right from the government (in the case of government land) or from a private owner (in the case of private land). In the case of open-cast mining, both rights are complimentary. Mineral rights sans surface right is meaningless.

In a competitive market scenario, a surface right owner would oblige a mining firm only when his/her reservation price matches the firm's willingness to pay. However, the market fails facilitate such smooth exchanges. Considering the geographical specificity of minerals and the national interest, the government intervenes to acquire land. Does the geographical specificity provide a rationale for designing special compensation packages for landowners? Should the land or surface right owner get a share of the mineral revenue? Or should landowners in mineral-rich areas only be compensated the same way as those in areas where other development projects are implemented?

A mining firm pays royalty to the owner of the mineral rights, which is the government, as well as fees and taxes. How can the surface right owner get a share of the profit generated from minerals owned by the government? Long (1995) points out that a passive landowner has no claim to the value added in extraction and processing but can exploit the lack of alternative mining sites. Landowners want higher payments than they would get from alternative uses of their land (a reservation price), but they will negotiate for the best deal above that amount. Landowners and mining firms will negotiate their respective shares of anticipated differential rents. Long defines differential rents as the profits over and above what is necessary to bring about mineral investment. These occur when some deposits being mined are cheaper to develop and mine than others. The highest-cost profitable mining venture will generate just sufficient net income to cover the mining firms' cost of capital and the lowest of all landowners' reservation prices. Any other deposits that are cheaper to locate, develop, mine, and process will accrue differential rents.

A landowner's share of these differential rents cannot be interpreted as an in situ value of the mineral reserve since at the margin that value is zero. If all deposits were equally costly to mine, landowners would only receive the reservation price for their lands. Thus, the simple concept of being paid for the value of minerals extracted is flawed. Landowners are paid what is necessary to induce them to forgo all other alternative land uses plus whatever differential rents they can bargain for (Long, 1995; p. 78).

In India, however, until now, no such mechanism exists. Weighing the merits and demerits of the mechanisms that do exist in the Indian context is beyond the scope of this chapter. Further, there is hardly any study that compares the social compliance of public and private firms. Given some similarities between CSR and social compliance, we can draw some inferences from the literature on the former.

A case study of British-Australian Rio Tinto Group, which is world's largest mining multinational, by Kapelus (2002) reveals that although private firms undertake steps to appear socially responsible, they in reality fail to implement their "idealistic principles and guidelines". Ghazali (2007) shows larger companies and companies in which the government is a substantial shareholder disclosed significantly more CSR information, while companies in which the executive directors held a high proportion of the equity (or owner-managed companies) disclosed significantly less CSR information. Literature on CSR points out a set of other indicators—the size of a firm, its age and its profitability—has a significant influence on social performance. A survey of the literature on the relationship between CSR and profitability by Aupperle, Carroll, and Hatfield (1985) could establish no direct link. Cochran and Wood (1984) demonstrate a high correlation between the average age of corporate assets and CSR ranking. After controlling for this factor, the correlation between CSR and financial performance remains, although it is somewhat weak.

Most of the literature on CSR comprises ex post analyses of firms' social performance. However, the issue of land acquisition and payment of compensation to tenants requires ex ante analysis. The findings of Szablowski (2002) are very much relevant to the present study. He uncovers inappropriate social impact assessment, low compensation to project-affected persons (PAPs) and breach of trust by Compania Minera Antamina (CMA) of Peru. He further points out that CMA violated procedures by not recruiting social experts and not involving the local community in carrying out impact assessment studies.

Herbert and Lahiri-Dutt (2004) discuss the failure of India's state-owned CIL, one of the largest coal producers of the world, to provide satisfactory compensation to the PAPs in Parej East in Jharkhand. During negotiations with landowners, CIL authorities promised to give jobs and substitute land to those affected by the coal mine. None of these assurances have been fulfilled in a satisfactory manner. Bhengara (1996) points out that even the baseline surveys on which rehabilitation was based were flawed. The baseline survey did not value the assets such as trees, wells and other non-movable assets, which are entitled to compensation under the Coal Bearing Areas (Acquisition and Development) Act, 1957 and the LAA. Similarly, compensation was not paid for the land and other common property resources over which the community had only customary rights but no legal record. No value was also accorded to the cultural and religious institutions of adivasis.

Serious economic analysis of involuntary land acquisition, displacement and resettlement are almost non-existent. Cernea (1999; p. 2150) points out that basic research on current population displacements is not carried out by economists anywhere. Development economics still has to respond in full to the challenge of analysing not only the economic dimensions of dislocation, but also the economic content of the other social costs of development.

The present study makes a modest attempt to bridge this gap by comparing the social compliance of public and private mining firms.

5.4. Ownership and Social Compliance: Theory

In this section, we discuss the theoretical factors that could contribute to a possible difference in the social compliance of public and private mining firms.

(i) The first point of difference between public and private firms is their primary objective. Public-sector firms, by and large, have the aim of maximising social welfare, sometimes even compromising on profit; whereas private firms solely focus on maximising profit. This would motivate public-sector firms to provide higher compensations than private-sector firms.

(ii) Public-sector firms have the advantage of using coercive powers to acquire private land. Private firms lack this and may have to tread the incentive route to ease the land acquisition process. As a result, private firms could end up providing a better compensation package than public-sector firms.

(iii) Public-sector firms may not face budget constraints in providing better compensation packages. Apart from higher monetary compensation, public-sector firms may provide more employment opportunities to landowners, even compromising on their profit. However, private firms would have limitations on providing both cash and employment compensation, and end up in parting with less than public-sector firms. Only very large private firms would be able to provide better cash compensation but they would still be reluctant to expand employment.

(iv)During land acquisition, a major demand everywhere is that the affected be given land elsewhere in exchange for the acquired land. Since the state owns all public land, only public-sector firms are in a position to facilitate such an exchange. Private firms will face much hardship to do so.

In extreme cases where landowners stick to their demand for land, private firms may not be able to acquire any private land.

(v) Given that the state can regulate land acquisition and compensation, the social compliance of public and private firms would vary depending on their bargaining power with the government authorities. The bargaining power of public firms is bound to be stronger than that of private firms. However, rent-seeking activities by private firms would reverse the scenario. Managers of public firms are left with little scope to bribe the regulatory agencies but those of private firms could do so to influence decisions in their favour.

(vi) The size of a firm largely determines its social compliance. Large firms would be in a position to provide better compensation than small firms. They would also be more concerned about their social reputation, for a good reputation not only keeps the market value of their shares high but also helps in their smooth operation. Delays of large projects due to public opposition seriously damage a firm's reputation and increase costs. So it makes sense for large firms to provide better compensation packages than small firms. If public-sector firms are larger than private-sector ones, their social compliance could be better.

(vii) The difference in the social compliance of public and private firms could also be explained by the theory of public choice and the principal-agent problem. The fixed salaries of managers in PSEs do not give them any incentive to bother about profits. Also, since political leaders tend to take the credit or discredit for the good or bad PSEs do, managers are not very bothered about the reputations of their firms and find it convenient to just follow the guidelines provided by legislators. As most citizens are largely uninformed about regulatory decisions, and lack incentives to become sufficiently informed to reward managers of PSEs who do not shirk, managers do not, or cannot, protect the broad regulatory interests of citizens. Therefore the social compliance of a PSE would be nothing more and nothing less than government policy. In the private sector, managers would be much more concerned about the reputation of their firms and the profits they make. Depending on its size and reputation, the manager of a private firm could plan to provide a better compensation package than even what the government advises. On the other side, private firms could influence politicians or regulators through rent seeking and keep the compensation low.

All these theoretical arguments provide us with no absolute answer to the question of whether public or private firms tend to provide better compensation and the issue is one on which empirical data has to be gathered.

5.5. Data and Methodology

5.5.1 Data Collection and Survey Area

To compare the social compliance of public and private mining firms, the study relies on primary data sources. Sixty-nine households that surrendered their land to mining firms between 1980 and 2008 were surveyed in Odisha. Two mineral-rich districts, Kendujhar and Jajpur, were selected for this and most of the households were in six villages in four of their panchayats. Detailed information on the sample has been presented in Table 5.1. Lists of affected households were obtained from the district collectorates by applying for them under the Right to Information Act, 2006. Of the 69 households, a large number had surrendered their land to public-sector mining firms and a few had surrendered their land to more than one mining firm. Thus, the 69 households were involved in a total of 86 land transfer cases. Complete information could not be obtained in two cases and our analysis is therefore confined to 84 cases of land transfer. Of these, 53 transfers were to public-sector firms and the remaining 31 to private-sector firms. The number of households, and the mining companies they transferred their land to, are as follows: 31 to Orissa Mining Corporation (OMC); 13 to Sirajjuddin Pvt. Ltd.; 2 to Rungta Minerals; 5 to Patnaik Minerals Pvt. Ltd.; 11 to FACOR; and 22 to the Industrial Development Corporation of Odisha (IDC).

Table 5.1: Selection of the Surveyed Villages

<table>
<tr><th>Districts (2)</th><th colspan="3">Kendujhar</th><th colspan="3">Jajpur</th></tr>
<tr><td>Blocks (2)</td><td colspan="3">Joda</td><td colspan="3">Sukinda</td></tr>
<tr><td>Panchayats (4)</td><td colspan="2">Balada</td><td>Bambari</td><td colspan="2">Kaliapani</td><td>Kansa</td></tr>
<tr><td>Villages (6)</td><td>Guruda</td><td>Balada</td><td>Palsa</td><td>Gurujang</td><td>Ostapal</td><td>Talangi</td></tr>
</table>

Before the survey, in November-December 2008, information on the land acquisition process and modes of negotiation and compensation were gathered through extensive discussions with the various stakeholders in the land transfer process. These included officials at many levels in the Department of the Directorate of Mines, Odisha, and some of the affected households. Keeping in mind all the information gathered, a questionnaire was designed in English and later translated into Oriya. After conducting a pilot survey of few households, necessary modifications were made to the questionnaire. The actual survey was conducted in April-May 2009. The interview schedules had questions related to the mode of land transfer (direct purchase, lease or acquisition by paying compensation), the process of transfer, the nature of compensation, the satisfaction of households with the compensation, and suggestions to improve the compensation package.

Table: 5.2 Education and Occupation Profile of the Surveyed Households

Education		Principal Occupation	
Level	% of total population	Occupation	% of total population
Illiterate	32.4	Public Sector Jobs	6.9
0-8	47.9	Private Sector	4.6
09-10	16	Agricultural labour	1.8
10-12	2.8	Own Farming	1.8
		Casual labour	1.6
		Total work participation	16.7

Total Population: 494

Source: Survey data

The villages in which the survey was carried out were mostly located in remote, hilly and dry areas. Most of the households surveyed belonged to the adivasi community. At the time of the survey, none of the villages had an all-weather road. All of them had at least one primary school, and very few households had any interest, and the ability, to educate their children beyond the primary stage. Barely a few children were matriculates and most of the villagers had little idea of what modern education was. Table 5.2 presents the education and occupation profile of the surveyed households. Only 16.7% of the surveyed population participated in paid jobs. Out of which around 11% worked in the organised sector (both public and private) and rest worked in the unorganised sectors. The mining firms sought the help of local politicians to ease the land acquisition process. In one village, a few households stated that their land records were deposited with the *pradhan* (village headman). To our surprise, the *pradhan* refused to talk to us. Consumption of alcohol is widespread and a large part of the compensation money has gone to country liquor vendors. Regular drinking, it is believed, prevents one from falling ill with malaria. Many of households had quit agriculture and by and large depended on income from working in the mines.

However, many people work as casual labourers at a low wage with no security of tenure.

5.5.2 Methodology

As stated earlier, when comparing the social compliance of public and private mining firms, we compare the compensation provided by the firms to the households that surrendered land to them. Mining firms compensated households directly in two ways—cash and employment. In our survey area, there were no cases of land-for-land compensation or house-for-house compensation. The rate of cash compensation (cash-for-land) was determined by the government regulatory agency. Therefore, it was the same for both public and private mining firms. Differences emanate from job compensation. Keeping this in view, a comparison of compensation is made by comparing the job compensations provided by public and private mining firms. It is hypothesised that public sector firms would be in a position to provide more jobs than the private firms. Similarly, we have also attempted to examine whether the mode of land acquisition and land category influence the job compensation. The statistical significance of the difference is tested through a logistic regression model. This can be written as below.

Model 5.1

$$\text{jobcomp} = f(\text{firmown}, \text{acqcat}, \text{landcat}) \qquad (5.1)$$

where "jobcomp" refers to whether any of the family members has got a job in the mining firm concerned as compensation. The responses are yes and no, coded as 1 and 0 respectively. Next, "firmown" indicates whether the mining firm is public or private; "acqcat" indicates the nature of land transfer, whether acquired paying compensation or on lease and "landcat" refers to the category of land transferred by a household –irrigated cultivable land, un-irrigated agricultural land, fallow land, and house plot. A few households transferred both un-irrigated agricultural and fallow land which has also been categorised as another group.

5.6. Results and Findings

5.6.1 General Observations

Mining firms in the studied areas took over tenanted land in three ways—by direct purchase, by paying compensation to the tenant, and on lease. Of the 82 valid cases of transfer, eight were by direct purchase, 31 were by paying compensation and the remaining 43 were taken on lease. Officially, it is believed that the mode of transfer was chosen after the negotiations between tenants and mining firms. Except in the cases of direct purchase of land, mining firms provided job compensations to many, though not all, of the households that surrendered their land. Compared to the households that transferred their land on lease, a higher proportion of those that surrendered land with compensation got jobs in mining firms. Of the 31 households that surrendered land after receiving compensation, 90% got job in the mining firm that acquired their land. But of the 43 households that leased out their land, only 28% got a job in the mining firm concerned. The category of land was immaterial in providing job compensation.

A job was the most preferred mode of compensation, followed by land and cash. Households that preferred jobs, land and cash as compensation were 49%, 19% and 8% respectively. The rest preferred to get cash and a job, a job and land, or all of them. Around 8% of the households demanded that they be given cash, a job and land. A large number of households that got a job in compensation feel that the job should be transferable to the next generation. In the case of joint families, the most pressing demand is that jobs be provided to all the members of the family. There have been disputes in such families over who should take up the job offered.

Raising the living standards of local communities, greater employment opportunities, and better economic and social infrastructure are the major promises made by all promoters of new projects while acquiring land. Nonetheless, the findings from our survey area tell a different story. 6 out of every 10 households were of the opinion that their economic condition had either remained unchanged (three) or worsened (three) after mining operations began. The environmental quality in the mining areas has deteriorated significantly. The air is heavy with dust due to the frequent movement of heavy vehicles. Dust from the mines can also be seen everywhere—in water, air, food, and so on. Although hundreds of heavy vehicles frequently run through the area damaging its makeshift roads, no attention has been paid to connecting the villages with proper roads.

However, every 6 out of 10 households felt that employment opportunities in the locality had improved because of the mining projects.

According to 58% of the households, education and health facilities and the supply of drinking water had improved after mining operations began. But many of the households surveyed expressed dissatisfaction over some of the facilities as well as the quality of service. For example, respondents noted that doctors visited the villages only once in a week. "Disease however does not know the schedule of the doctor," pointed out one respondent. Major facilities are limited to the households that surrendered their land. Drinking water is provided by tankers. Mining firms have also placed water tanks in a few places. Villagers, however, complained that water was not provided regularly. That the provision of drinking water is perfunctory is evident from the careless manner in which water tanks are situated here and there.

Every 6 out of 10 households were of the view that no perceivable improvement in household asset holdings had occurred and 1 out of 10 households opined that it had declined (one) after mining operations began. In possession of domestic animals, 5 out of 10 households reported no change and rest 4 out of 10 households said there was a decline. This more or less nullified the positive response on availability of jobs. Most households have got either casual jobs or very low-grade jobs. Mining firms would say that this is because individuals lack skills. However, mining firms have taken no initiative to build capacity among the villagers. Most households felt that a job had not compensated for the loss of land as well as common pool resources. This could explain why many households responded that there had been no perceivable improvement in their economic condition or that it had worsened.

As mentioned earlier, a major paradox is that mineral resources are very often beneath forest land.[3] Exploitation of minerals involves massive deforestation, which results in environmental degradation and the impoverishment of the people depending on forests for their livelihood. More than half the households surveyed reported that the public land they had had access to was now in the hands of the mining firms. The public land was used by households for various purposes, including gathering firewood, collecting non-timber forest products, grazing animals and agriculture. Mining operations had fully denuded the nearby forests and grazing lands. This could partly explain the fall in the possession of domestic animals.

---

[3] For example, see CSE (2008; p.7).

The degradation of common pool resources had equally affected both households that surrendered their land and those that still have their land. The compensation package provided by the mining firms did not make any allowance for this. An unequivocal complaint of the people and activists in the area was that mining operations were depleting groundwater and drying up the topsoil. This had caused a serious shortage of drinking water and a fall in agricultural productivity. Moreover, run-off rain water from the mountain-like overburden dumps carried poisonous silt to the fields. Springs in the locality, on which villagers earlier depended for drinking water and irrigation, had either been polluted or had dried up after the mining began. This had seriously affected agricultural productivity and many households said they had stopped cultivation.

The compensation package had not made 80% of the households happy but only 14% of them said that they had approached the court to secure a better deal. A public hearing in the locality before land is acquired has only recently emerged as an official requirement. The surveyed households that had surrendered their land in the 1980s or 1990s reported that mining firms held meetings in their villages. Local political leaders, who now control labour unions in the mines are hand in glove with the mining firms, played an important role in such meetings where the villagers were mute spectators. A group of individuals, who claimed to have no political allegiance, said that the influence of one of the local political leaders on the labourers in the mines and the adivasis who had surrendered land was such that they would not express their views on the compensation package or speak a single word against any mining firm.

5.6.1.1 Predatory Land Acquisition

A few households that initially refused to surrender their land were later forced to do so due to a predatory policy followed by the mining firms. Control over land was secured by the mining firms wherever possible and operations began. As pointed out earlier, mining seriously affects the productivity of adjacent lands. In view of the shrinking return from their land, many households had no alternative but to surrender their land to the mining firms. And now, there are many others in the queue.

5.6.2 Comparative Analysis

Having discussed the general findings on the land acquisition process, the payment of compensation and the impact of mining, this section will look into the differences between public and private mining firms. Before carrying out the survey of households, attempts were made to hold discussions with the officials of the mining firms concerned to comprehend their policy on land acquisition and compensation. The officials showed little interest in disclosing any information to us. The general manager of a public-sector mining firm became furious on being requested to share information on the compensation paid to the villagers. "Our company might not be giving the best compensation package. Why should we share the information? We do not want to invite any trouble by sharing such sensitive information with you," he said. This went against the popular notion that public-sector firms offer better compensation.

Given the reluctance of both public- and private-sector mining firms to share any information, a comparison of public and private mining firms has been made through assessing the job offers made to the households; examining the public facilities provided by the mining firms in the locality; and gauging the overall satisfaction of the households with the compensation they received. Apart from collecting information on their compensation, we asked households whether they were satisfied with it or not. Which type of firm—public or private—did the household believe provide better compensation? For this analysis, only the households that surrendered their land after receiving compensation or on lease have been taken into consideration. Here, of course we need to take a caveat that there could be some subjectivity involved in this answer. Similarly, this perception could also change over time. For example, households showing satisfaction over the compensation today may opine to be dissatisfied tomorrow due to rise in expectation. Nevertheless, assuming the same level of subjectivity for both public and private firms we can make a comparison.

After dropping the land transfers through direct purchase, we have 75 cases where land was transferred on lease or after receiving compensation. 41 of the 52 (around 79%) households involved in the land transfers to public-sector mining firms, expressed dissatisfaction over the compensation. Likewise, households involved in 22 (around 96%) of the 23 transfers to private mining firms, said they were dissatisfied with the compensation.

Forty households from those involved in the 75 land transfer cases got jobs in the mining firm concerned as compensation. Households involved in 29 (around 56%) of the 52 land transfers (either on lease or after receiving compensation) to public-sector mining firms got jobs as a part of the compensation package. Households involved in 11 (around 50%) of the 23 land transfers (either on lease or after receiving compensation) to private-sector mining firms also got jobs. It can be inferred from this that the public-sector mining firms provided more jobs than the private-sector ones. For ascertaining the statistical significance of the difference between the public and private mining firms in providing jobs as compensation, we have undertaken a logistic regression analysis (equation -5.1).

We present the results of the regression analysis to determine whether firm ownership made any differences to the compensation provided to the households as jobs. The results of the logistic regression analysis are presented in Table 5.3. The outcome variable "jobcomp" is regressed on the explanatory variable "firmown" and the control variables "acqcat" and "landcat". To differentiate the effects of each explanatory variable, we show the results with four models. Model 5.1.1 takes "firmown" as the only explanatory variable. The "p" value for the odds ratio conveys that firm ownership does not really make any difference in getting job compensation. In Model 5.1.2 "acqcat" has been taken as the only explanatory variable and the "p" value for the odds ratio is significant at the 1% level. This allows us to infer that households that transferred their land by receiving compensation had a greater chance of getting a job than those that leased out their land. In Model 5.1.3 the odds ratio for the explanatory variable "landcat" remains significant for two categories—households that transferred fallow land and those that transferred both unirrigated agricultural and fallow land. The odds ratios for these two categories of households convey that their probability of getting a job as compensation was less than that of households that transferred irrigated agricultural land. Model 5.1.4, where we have entered all three explanatory variables, allows us to infer whether the ownership variable tells the same story after controlling for other variables. The odds ratios for the variable "firmown" in Model 5.1.1 and 5.1.4 are insignificant.

From this we can safely conclude that there was no difference between the public and private mining firms in providing jobs as compensation. In other words, firm ownership made no difference to the social compliance. In the final model (5.1.4), however, it is important to highlight that households that surrendered their land by receiving compensation had more probability of getting jobs as compensation than those who leased out their land. More interestingly, in the final model, the land category becomes insignificant in determining jobs as compensation.

Apart from the quantification of the difference in job compensation we have gathered the opinion of households on the overall compensation (which includes the direct compensation made to the households and indirect compensation through public facilities like drinking water, health services etc) of the public and private mining firms. Households were asked whether private or public mining firms provided better compensation. The responses were one of these four—public, private, indifferent and do not know. Of the 50 households that transferred land to public-sector mining firms, 21(42%) believed that public firms provided better compensation than their private counterparts. Of the remaining households, 26% believed that private firms provided better compensation, 18% did not see any difference between public and private firms, and 14% did not provide any opinion.

Table 5.3: Logistic Regression Analysis

Model-5.1: Outcome Variable: jobcomp: yes = 1, No = 0 Odds Ratios				
Explanatory Variables	**Model 5.1.1**	**Model 5.1.2**	**Model 5.1.3**	**Model 5.1.4**
firmown Private	0.671 (0.421)			0.678 (0.632)
acqcat@ Withcompensation		24.111* (0.000)		24.685* (0.007)
landcat Unirrigated land			0.335 (0.139)	2.422 (0.463)
Fallow land			0.016* (0.001)	0.223 (0.345)
Both irrigated & unirrigated land			0.035 (0.008)	0.650 (0.794)
N	76	74	67	66
Pseudo R2	0.0062	0.3082	0.2739	0.3932

Notes:
Values in parentheses are "p" values reflecting the statistical significance of the relative risk ratios.
In variable "firmown", the public sector is taken as the reference category.
In variable "acqcat", direct purchase is the reference category.
In variable "landcat", irrigated land is the reference category.
@ implies acquisition, the category "lease" is dropped because of colinearity.
* significant at 1% level

Of the 13 households that transferred land to private mining firms, a large section (31%) believed that public mining firms provided better compensation. Thus, in aggregate, 40% of the households believed that public mining firms provided better compensation than private ones, 25% felt private mining firms provided better compensation, 19% did not perceive any difference between the two and 16% did not give any opinion.

From the opinion of households on the superiority of public or private mining firms in providing compensation, we do not derive any significant conclusion. Although, in aggregate, 40% of the households said that public-sector mining firms provided better compensation, 35% (19% indifferent and 16% with no answer) did not say so. And 25% of the households felt the compensation provided by private mining firms was better. This inconclusiveness can be read in tandem with the finding that a majority of the households (84%) were not satisfied with their compensation package, no matter who had provided it.

5.7 Discussion

It is imperative to read the results of our analysis against the backdrop of the theoretical propositions made earlier. Although the present study did not originally intend to compare the performances of public and private mining firms against any idealistic compensation package, a few inferences can be logically made, which could improve the current situation. The dissatisfaction of a majority of households with the compensation reflects the shortcomings of the compensation policy in force now. Our study reveals that a job remains the most preferred compensation for involuntary land transfer. For transferring the land, most households are given cash and a job as compensation. Some of them receive one of these two. In our survey area, there were no cases of land-for-land compensation. The rates for cash compensation of different categories of land was determined by government agencies and our household survey made it clear that both private and public mining firms paid the amount stipulated by them. Therefore we focus on jobs as compensation to examine the difference between what public and private mining firms offered. Though from a simple comparison, the percentage of households that received jobs by transferring their land shows a moderate difference. Of the households that transferred their land to public-sector mining firms, 56% got jobs as compensation; whereas only 50% of the households that transferred their land to private-sector mining firms got jobs. However, the difference between the public and private sectors in providing jobs as compensation is not significant in our regression exercise (Model 5.1). Thus, it would be pertinent to conclude that there is no significant difference in the social compliance of public and private mining firms.

The overall dissatisfaction with the compensation packages received could be seen as a policy and regulatory failure. As pointed out in most of the literature on displacement, the incumbent compensation policy takes into account only the direct economic losses of the PAPs. Indirect economic losses caused by environmental degradation and the loss of common pool resources are seldom compensated for. For marginal landowners and adivasis, such costs are enormous. The households surveyed reported serious losses in productivity in agriculture, forest resources, pastures and public land, which were often used for cultivation. These losses have not been compensated. Households and mining firms agreed that the compensation package would be determined only on the basis of the area of land surrendered, which meant making up only for direct economic losses. Households that did not surrender land but have been affected by the environmental and social damages caused by mining do not even come into the picture.

5.8. Conclusion

Involuntary land acquisition remains a highly contentious issue in most of the developing world. In this chapter, we compared the overall social compliance of public and private mining firms. The theoretical arguments for explaining possible differences in the compensations provided by public- and private-sector firms do not provide a definite conclusion, making this an empirical issue. To explore the possibility of there being any difference, 69 households (making up 84 land transfer cases) that had surrendered their land to mining firms were surveyed in two mineral-rich districts of Odisha.

The study shows that a majority of households were dissatisfied with the compensation they received from public and private mining firms. From a simple comparison, we find a moderate difference in the percentage of households that received jobs by transferring their land to public and private mining firms. Of the households that transferred their land to public-sector mining firms, 56% received jobs as compensation; whereas only 50% of those that transferred their land to private-sector mining firms got jobs. Nonetheless, the difference between the public and private sectors in providing jobs as compensation is not significant in our regression analysis.

Discussions with households in the mining region that had not surrendered their land to mining firms pointed to the uncompensated damage caused by the acquisition of common pool resources and the negative externalities of mining in the locality. This was uniform in the case of both public and private mining firms. As pointed out in most of the literature on displacement, the current compensation policy takes into account only the direct economic losses of the PAPs. Indirect economic losses caused by environmental degradation and the disruption of longstanding social arrangements are seldom compensated for. Keeping in view the growing demand for minerals, the study suggests there be a paradigm shift in the incumbent compensation policy.

CHAPTER SIX

SUMMARY AND CONCLUSION

6.1 Research Issues Addressed in the Study

From the early 1990s, most of the developing nations have opened up their mining sector to the participation of both domestic and foreign private firms with a view to raising the production and boosting the productivity of the mining industry. The participation of private players is expected to increase production and overall productivity of the mining sector through direct and indirect effects. Under direct effects, it is presumed that private firms are more efficient than their public counterparts and their participation will bring in more capital, better technology and superior managerial skills, thus raising the overall productivity of the sector. Under indirect effects, the participation of private players is supposed to increase competition in the sector, while the spread of superior technology leads to an increase in overall productivity. The thinking is that in the face of competitive pressure, inefficient public firms will attempt to raise their productivity levels, at least to be at par with those of efficient private firms, and this will increase the overall productivity of the mining industry.

Liberalisation of mining industry also raises serious apprehension of accentuating environmental degradation and the marginalisation of local communities. Mining industry is considered to be one of the most polluting industries, especially mining of chromites, coal and uranium. The basic textbook knowledge of public economics and environmental economics tells us that private firms only focus on the maximisation of profit and seldom bother about the environmental health of the mining periphery. The greater participation of private firms in the mining industry therefore amplifies the apprehension of increasing environmental damage.

Mineral extraction also causes displacement of local communities, mostly the indigenous communities who are marginalised both economically and socially (Downing, 2003). There have been serious allegations on the payment of no compensation or much less than the desired level of compensations to the project affected households

(Fernades et al., 1997). But is there any difference between the public and private mining firms in providing compensation? Drawing arguments on the line of environmental performance of public and private mining firms we can hypothesise that public firms will provide better compensation (better social compliance) compared to the private mining firms. However, better economic performance of the private mining firms would enable them to provide better compensation.

Keeping in view the above three issues (productivity, environmental performance and social compliance), the present study sought to examine the following questions: i) Are private mining firms more productive (in extraction) than their public-sector counterparts? ii) do public-sector mining firms comply with environmental regulations better than their private counterparts? and iii) do public-sector mining firms provide better compensation than private firms for acquiring private land?

For examining the above research questions, the study focused on the Indian mining industry. To provide a broader canvas, the policy changes that have taken place within the mining industry in the post-liberalisation period were described. The share of mining and quarrying activities in the NSDP of the states and the GDP of the country were then analysed to quantify the contributions of the mining industry to the state and national economies. An analysis was also carried out on the share of royalty in the total revenue receipts of the states. From these analyses, it emerged that although the M&Q sector has a negligible share (1.98% in 2007-08) in the overall GDP of the country, it has a significant share in the NSDP of Chhattisgarh, Jharkhand, Meghalaya and Odisha. In 2007-08, Chhattisgarh had the highest share from the M&Q sector in its NSDP, 14.43%. All other states recorded less than 10% share of the M&Q sector in their NSDP. Jharkhand ranked second in the contribution of M&Q to its NSDP (8.66%), followed by Meghalaya (8.35%), Odisha (6.39%), Goa (4.13%), Madhya Pradesh (3.90 %), Andhra Pradesh (3.56%) and Assam (3.29%). In the contribution of the M&Q sector to revenue, Jharkhand, Chhattisgarh and Odisha ranked among the top three. Jharkhand earned the most from royalty paid by mining firms, followed by Andhra Pradesh, Madhya Pradesh, Chhattisgarh and Odisha. In percentage contribution to the total revenue receipts, Jharkhand headed the list with 12.54 %, followed by Chhattisgarh, Odisha and Rajasthan with share of 9.31%, 5.77% and 3.37 % respectively.

These positive contributions aside, the mining industry has had a severe negative effect on environmental health and the social fabric of the periphery. Vast stretches of forests have been cleared to give way to mining operations (CSE, 2008). Due to the popularity of open-cast mining, a large number of adivasis have been displaced. Many of them have either not been compensated or inadequately compensated for their economic and social losses (Fernades et al., 1997). One estimate by the government puts the total forest land diverted for mining between 1980 and 2005 at 95,003 ha. Other sources point to a much higher figure. Based on information available from various sources, including the Union Ministry of Environment and Forests (MoEF), the total forest land diverted for mining in India in the 1980-2005 period has been estimated to be as high as 164,610 ha. Even this figure would be higher if it took into account the forest land diverted before 1980 when many coal mines took over vast areas of land—mostly forests.

Mining projects have caused massive displacement of the adivasi population. No systematic data is available on the number of people who have been displaced because of mining projects. The most cited data, provided by Fernades et. al. (1997), reveal that between 1950 and 1991, mining projects have displaced around 25.5 lakh people. Significantly, not even 25% of these displaced people have been resettled. These figures only represent the population who were moved out of their lands but do not include the thousands who were dependent on the land for their livelihoods, or those whose lives were affected due to disruption of water tables, dumping of overburden on fertile agricultural land and destruction of forests. The worst victims of mining projects have been adivasis. Of the total population displaced by various development projects, about 41% are adivasis. In the case of mining projects, about 52% of those displaced are adivasis.

6.2 Methodology Adopted for the Study

Before empirically verifying the three research questions mentioned, we analysed, theoretically, the efficiency, equity and environmental implications of Indian mining laws (Chapter 2). From this analysis, we drew the conclusion that state governments should possess the ownership rights over minerals while the Central government retained the regulatory authority for ensuring the free flow of minerals across states and sectors, safeguarding the environment, securing social welfare, and achieving sustainable development without sacrificing strategic concerns. Nonetheless, local governments can play a crucial role in monitoring mineral production.

The theoretical analysis also provided the rationale for vesting mineral ownership rights in the state governments, and granting permission for extraction to both public and private firms to foster competition and raise productivity of the sector. The review of theoretical literature on the main research questions—whether public and private mining firms perform differently in productivity efficiency, environmental compliance and social compliance—did not provide us with any single conclusion and they were therefore open for empirical investigation.

For examining the first research question, whether firm ownership has any bearing on the productivity of mining firms operating in India, the study used the firm-level data provided by the CMIE in its electronic data base, Prowess. Firm–level data at two-digit NIC has been used for the period from 1988-89 to 2005-06. Using an unbalanced panel data set, the study compared (i) TFP levels and the growth rates of public and private firms in the four sectors of the Indian mining industry—metallic, non-metallic, coal and petroleum; and (ii) TFP levels and growth rates in the pre-liberalisation and post-liberalisation periods.

For examining the second research question, whether firm ownership differentiates the environmental performance of mining firms operating in India, the study focussed on the chromite mining industry. First, a comparison was made, separately, on four indicators—i) quality of mine drainage water; ii) management of overburden; iii) ambient air quality; and iv) quality of drinking water. In the second step, an aggregate environmental performance measure named the multidimensional environmental defiance index was constructed and the environmental performances of public and private firms were compared through a permutation test. On the methodology front, the study offered a new measure to assess the environmental performance in a multidimensional[1] framework, which satisfies the monotonicity property with respect to breadth and intensity of pollution.

For examining the third research question, whether firm ownership differentiates the social compliance of the mining firms, we undertook a case study of the Odisha mining industry. The study used primary data collected from a survey of 69 households that had surrendered their land to public and private mining firms. Through the survey, data were collected on the mode of land transfer (direct purchase, lease and through acquisition by paying compensation), the process of transfer, the nature of compensation, the satisfaction of households with the compensation received, and suggestions to make the compensation package better.

The concept "social compliance" has been defined as the direct and indirect compensation made to, and the facilities provided, for an

[1] The methodology has been adopted from the multidimensional poverty literature (Alkire and Foster, 2008)

individual or the community for the direct and indirect economic, social and environmental losses suffered due to the involuntary transfer of property, displacement and damage to the local natural resource base. Mining firms compensated the households in our survey area directly in two ways—cash payment and employment. There were no cases of land-for-land compensation or house-for-house compensation in this area. Apart from direct compensation to the households, mining firms make public provisions for the communities like supplying drinking water, and improving health and education facilities. Therefore, the study compared both types of compensation.

6.3 Findings of the Study

A comparison of TFP levels of public and private mining firms showed the superiority of private firms in three sectors—metallic, non-metallic and coal—during the entire period of analysis. In the petroleum sector, private firms initially outperformed public firms but eventually TFP levels of public firms exceeded that of private firms. The productivity gap between public and private firms remained highest in the non-metallic sector. Private firms in this sector were almost two times more productive than their public counterparts. Similarly, private firms in the metallic and coal sectors were one and half times more productive than their public counterparts. Competitive pressure through liberalisation has not been able to bridge the productivity gap between public and private firms.

Liberalisation of the mining industry has brought about increased private participation. Consequently, the share of the public sector in the total value of output declined from 91.19% in 1988-89 to 74.61% in 2004-05. Private participation is highest in the non-metallic sector. For example, in 2003-04, the share of the private sector in the value of total limestone production was 93.6%. Between 1988-89 and 2005-06, TFP levels of the mining industry rose by one and half times and, on an average, TFP grew at the rate of 2.52% per annum. However, inter-temporal comparison of TFP growth shows differences in performance in three sub-periods. In the pre-liberalisation period, TFP of the mining industry grew faster than in the post-liberalisation period. While TFP in the pre-liberalisation period grew at the rate of around 4.5% per annum, it slowed down in the first phase of liberalisation, growing only at the rate of 0.94% per annum. However, the momentum of growth bounced back in the second phase of liberalisation and TFP grew at around 3%. On average, TFP growth in the liberalised period (1.91%) was lower compared to the pre-liberalisation period (4.49%).

A comparison between the environmental performances of public and private mining firms in four indicators separately gives inconclusive results. The construction of a single environmental performance measure in the form of the multidimensional environmental defiance index facilitated a better comparison between public and private mining firms. With the help of both unidimensional and multidimensional indices, the study did not find out any significant difference in the environmental performances of public and private mining firms.

The study revealed that a majority of households were dissatisfied with the compensation received from both public and private mining firms. A simple comparison of the percentage of households that received jobs by transferring their land to public and private mining firms showed a moderate difference. Of the households that transferred their land to public-sector mining firms, 56% received jobs as compensation; whereas only 50% of the households that transferred their land to private sector mining firms received jobs as compensation. Nonetheless, the difference between the public and private sectors in providing jobs as compensation was not significant in our regression exercise. Therefore, the study concludes that there is no significant difference between the social compliance of public and private mining firms.

The overall dissatisfaction with the compensation packages could be viewed as a policy and regulatory failure. As pointed out in most of the literature on displacement, the current compensation policy takes into account only direct economic losses of PAPs. Indirect economic losses caused by the environmental degradation and damage to the social fabric are seldom compensated for. For marginal land-owners and adivasis, such costs are enormous. The surveyed households reported serious productivity loss in agriculture, loss of forest resources, and loss of pastures and public land, which was often used for agriculture. These losses of households were not compensated for. Households and mining firms reported that the compensation package was determined only on the basis of the total area of land surrendered; hence only the direct economic loss caused to the tenants. Households that did not surrender their land, but were affected by environmental and social damages, were not compensated. Discussions in the mining region with households that had not surrendered their land revealed the uncompensated damage that had resulted from the acquisition of common pool resources and the negative externalities of mining operations. This remained the same for both public-sector and private-sector mining firms. Keeping in view the growing demand for minerals, the study suggests for a paradigm shift that has to take place in the current compensation policy.

6.4 Policy Implications

In this study, we intended to examine the veracity of three common notions. They are, i) whether liberalisation of the mining industry will raise the output and overall productivity level; ii) whether more participation of private firms will accentuate environmental damage; and iii) whether more participation of private firms will aggravate social damage. With these questions in mind, we compared the performances of public and private mining firms in Indian context. Therefore, from our empirical analysis, we can draw some policy implications. Nevertheless, we must acknowledge that the policy inference drawn from this study should only be viewed as indicative for the simple reason that it is not too comprehensive. The basic goal of this study has been to develop and apply the most suitable analytical and methodological frameworks to examine some popular notions on the performances of public- and private-sector mining firms.

The study establishes that private-sector mining firms are significantly more productive than public-sector mining firms. On environmental and social compliance, there is no significant difference between public and private mining firms. Both public and private mining firms have failed equally in environmental and social compliance. Nonetheless, we shall refrain from providing any specific conclusion on whether liberalisation of the mining industry will have a positive or negative impact on aggregate social welfare. To address this question, we need to examine several other questions like the revenue implications of privatisation and the resource allocation of the government. The redistributive policy of the government in areas such as taxation and public expenditure in the mining areas will have a large bearing on social welfare. From reports of rampant corruption indulged in by private mining firms, one would have serious doubts about the merits of productivity gains from more private-sector participation.

The non-compliance of both public and private mining firms with environmental regulations and social obligations points to the palpable failure of the government regulatory mechanism. Therefore, further attempts should be made to explore the causes of regulation failure so that all loopholes can be plugged.

Similarly, further research needs to be undertaken to explore the determinants of differences in the productivity of public and private mining firms. In this regard, a close enquiry into the rise in productivity of public-sector petroleum-extracting firms may provide some useful insights.

6.4 Limitations of the Study

In spite of the significant contribution, the present study makes towards understanding the differences between the performances of public- and private-sector mining firms in productivity, environmental compliance and social compliance, we do acknowledge a few limitations. However, for several reasons, these limitations of the study are beyond the scope of easy rectification.

The study faced serious constraints in the availability of data. A major problem related to data on the mining sector is the regulated price in the coal and petroleum sectors. Nonetheless, the significance of the study is not undermined because we refrain from comparing the productivity across sectors. Rather, the study strictly focuses on a comparison of public and private firms within the same sector. Assuming that the value of output produced by both public- and private-sector firms are expressed in the same unit, the significance of the comparison is not diminished.

Within productivity estimation, the measurement of capital and the measures to address the endogeniety problem remain highly disputed (Griliches and Mairesse, 1998; OECD, 2001; OECD, 2001a). One could only find solace by acknowledging that all the estimation methodologies have one problem or the other.

In the comparative study of the environmental performance, there are serious constraints in the data available. The study relied on the data of one mining sector (chromite) and carried out its analysis with the help of data for 10 mining firms. Although, on the methodological front, the problem of comparison using a small sample size has been addressed by undertaking a permutation test, the other problems associated with a small sample remain. Data on the environmental performance of a higher number of firms will have definite merits compared to a small data set.

The study faced similar problems related to data in the comparison of social compliance as well. Data on social compliance has been gathered through interviews with the heads of the households that surrendered their land to the mining firms. Due to the large gap in time between land acquisition and compensation payment, and our survey, there is the threat of memory loss. This was clearly observed in the responses on the land acquisition process. However, to keep the analysis more objective, the study relied on definite variables, namely, job compensation. One could expect a certain degree of subjectivity in the responses of households' satisfaction with the overall compensation (direct and public facilities), and the main inferences of our study have been drawn primarily from the objective variables.

6.5 Issues for Further Research

To arrive at the larger welfare implications of liberalisation in the mining industry, a large number of other questions—such as the revenue implications of extractive firm ownership and redistributive mechanism of the government—need to be addressed before drawing any definite conclusion. In the following section, we outline a few research questions which need further investigations.

6.5.1 Revenue Implications

The government's revenue collection would vary because of two primary reasons—(i) the ownership of mining firms and (ii) Centre-state relations. In the first part, we shall discuss the differential revenue implications due to ownership. Next, we shall elaborate on the revenue implications of Centre-state relationship.

6.5.1.1 Firm Ownership and Revenue Implications

Ownership of mining firms by the private- and public-sector has different revenue implications. In the case of government ownership of mining firms, their entire profits would go to the public exchequer, whereas in the case of private ownership, the government can only collect royalty, various fees and taxes. However, the inefficiency of public-sector mining firms can sometimes lead to losses and hence a net loss to the public exchequer. On the other side, private-sector firms can evade taxes by several means. Under-reporting the quantity of mineral extracted or sold, under-reporting the price[2] (in the case of ad valorem tax) and bribing regulatory officials are some of these. Anecdotal evidence (discussions with mining officials) shows that private firms under-report the price of minerals (around 40% of the actual price) to the Indian Bureau of Mines to evade tax. In response to this illicit practice, the government imposes tax on a 10% higher price than the reported price. Similarly, due to weak monitoring, private mining firms under-report the quantity of minerals exploited. Recent reports of mining scams in Andhra Pradesh, Karnataka and Odisha indicate both regulatory failure and the rent-seeking activities of private mining firms.

The government could think of retaining the ownership of mining firms and undertaking adequate measures to improve their efficiency. One

[2] See Business Standard (2011)

way of increasing the efficiency of public-sector firms would be granting them more autonomy and linking the salaries of managers to profits. A recent move of the United Progressive Alliance Government to provide more autonomy to public-sector firms by categorising them as "Maharatnas" is noteworthy in this respect. The Indian experience shows that state-owned (both state and Central government) mining firms have reported reasonable profit levels. Therefore the issue needs systematic investigation before any definite conclusions are made.

6.5.1.2 Centre-State Relations and Revenue Implications

This study has argued for the state ownership of mineral resources. It has also been pointed out that Central government regulates mineral production by giving the final approval and determining royalty rates. However, due to faulty policy measures, royalty has been a major cause of revenue loss. The MMRD Act 1957 empowers the Central government to determine rates of royalty for different minerals and these are to be uniformly applicable in all states. The rates are revised by the Central government after a specific time period. To minimise the uncertainty due to revision of royalty, the MMDR Act 1957 provides for revising the rates once in three years. The Central government sets up a study group comprising representatives of state governments and the mining industry to revise royalty rates. Past experience shows that the royalty rates have not been revised in a timely manner and caused huge revenue losses to the state exchequers. From time to time, various states have attempted to collect extra revenue by levying different cesses and fees. However this has been contested in the apex court. In this context, it is imperative to understand the legitimacy of a uniform royalty rate across states. Das (2009) provides a detailed account of the faulty regulatory mechanism, which impedes the revenue collection of state governments, and calls for providing enough elbow room to state government to determine the royalty rates. Nonetheless the issue needs further investigation.

6.5.2 Determinants of performance difference

The study has attempted to explore whether there are differences between the performances of public and private mining firms but has not been able to provide any definite reason for such differences. Further studies should be undertaken for explaining such differences. What are the possible sources of such difference—management, technology, or capital constraint? The sources of differences would be different for different indicators—productivity, environmental performance and social compliance. Systematic investigations on these issues should be carried out to explore the causes of such failures.

BIBLIOGRAPHY

Alchian, A. A. (1965). 'Some economics of property rights', *Il Politico*, 30, pp. 816-829.

Abramovitz, Moses (1956). 'Resource and Output Trends in the United States since 1870', *American Economic Review*, vol. 46(2), pp. 5-23.

Agrawal, Neeraj (2007). 'How Green Was My Mountain', in (R. Kalshian, ed), *Caterpillar and the Mahua Flower: Tremors in India's mining fields*, New Delhi: Panos South Asia.

Akanni, O. P. (2007). 'Oil Wealth and Economic Growth in Oil Exporting African Countries', (170), Technical report, African Economic Research Consortium, Nairobi.

Akabzaa, Thomas and Abdulai Darimani (2001). 'Impact of Mining Sector Investment in Ghana: A Study of The Tarkwa Mining Region'. Draft report prepared for SAPRI. January 20.

Alkire, Sabina and James Foster (2008). 'Counting and Multidimensional Poverty Measurement'. OPHI working paper series.

Andersen, J. J. and S. Aslaksen (2008). 'Constitutions and the resource curse', *Journal of Development Economics,* vol. 87, pp. 227–246.

Arellano, Manuel and Stephen, Bond, (1991). 'Some tests of specification for panel data: monte carlo evidence and an application to employment equations', *Review of Economic Studies*, vol. 58(2), pp. 277-97.

Aupperle, K. E., A. B. Carroll, And J. D. Hatfield (1985). 'An Empirical Examination of the Relationship between Corporate Social Responsibility and Profitability', *The Academy of Management Journal*, vol. 28(2), pp. 446–463.

Auty, R. M. (2003). *Sustaining Development in Mineral Economies: The Resource Curse Thesis*, New York: Routledge.

Aydin, Hamit and John E. Tilton (2000). 'Mineral endowment, labour productivity, and comparative advantage in mining'. *Resource and Energy Economics.* 22, pp. 281–293.

Baland, J M and J. P. Platteau, (1996). *Halting Degradation of Natural Resources: Is There a Role for Rural Communities,* FAO, Rome: Oxford University Press.

Bartelsman, Eric J and Mark Doms (2000). 'Understanding productivity: lessons from longitudinal micro data'. *Journal of Economic Literature.* vol. 38, pp. 569-594.

Baumol, W. J. and W. E. Oates (1988). *The Theory of Environmental Policy*. New York: Cambridge University Press.

Bhattacharya, M. (2007). 'Nandigram and the question of development', *Economic and Political Weekly*, pp. 1895–1899.

Bhengara, R. (1996). 'Coal mining displacement', *Economic and Political Weekly,* 31(11), pp. 647-649.

Biesebroeck, Van, Johannes (2007). 'Robustness of productivity estimates', *Journal of Industrial Economics*, vol. 55(3), pp. 529-569.

Blundell, Richard and Stephen Bond (1998). 'Initial conditions and moment restrictions in dynamic panel data models', *Journal of Econometrics*. 87, pp. 115-143.

Boardman, Anthony e. and Aidan R.Vining (1989). 'Ownership and performance in competitive environments: a comparison of the performance of private, mixed, and state-owned enterprises', *Journal of Law and Economics*, vol. 32(1), pp. 1-33.

Brunnschweiler, C. N. (2008). 'Cursing the blessings? Natural resource abundance, institutions, and economic growth', *World Development* vol. 36(3), pp. 399–419.

Business Standard (2011). 'Government finds 17 miners undercutting iron ores prices'. http://www.business-standard.com/india/news/govt-finds-17-miners-undercutting-iron-ores-prices/443138/. July 19. Retrieved on 19th July, 2011.

Cavalcanti, T. V. d. V. Mohaddes, K. and M. Raissi (2009). 'Growth, Development and Natural Resources: New Evidence Using a Heterogeneous Panel Analysis' (CWPE 0946), Technical report, University of Cambridge.

Caves, Douglas W. and Laurits R. Christensen (1980). 'The relative efficiency of public and private firms in a competitive environment: the case of Canadian railroads', *The Journal of Political Economy*, vol. 88(5), pp. 958-976.

Centre for Science and Environment (2008). *Rich Lands Poor People: Is 'Sustainable' Mining Possible*?, New Delhi: Centre for Science and Environment.

Cernea, M. M. (1999). 'Why economic analysis is essential to resettlement a sociologists view', *Economic and Political Weekly*, vol. 34(31), pp.2149–2158.

—. (2003). 'For a new economics of resettlement: a sociological critique of the compensation principle', *International Social Science Journal,* (175).

Chakravorty, S L (2001). 'Artisanal and Small-scale Mining in India'. Report commissioned by the Mining, Minerals and Sustainable

Development project of International Institute for Environment and Development. No.78.

Chamberlain, G. (1982). 'Multivariate regression models for panel data', *Journal of Econometrics,* 18, pp. 5-46.

Chandra, N. K. (2008). 'TATA motors in Singur: a step towards industrialisation or pauperisation?', *Economic and Political Weekly*, vol. 43(50), pp. 36–51.

Charles, R. Hulten, (1979). 'On the importance of productivity change', *American Economic Review*, vol. 69(1), pp. 126-136.

Cochran, P. L., and R. A. Wood (1984). 'Corporate Social Responsibility and Financial Performance', *The Academy of Management Journal*, 27(1), pp. 42–56.

Commision on Sustainable Development (2000). 'Report on the eighth session'. Economic and Social Council, United Nations, New York. http://www.un.org/documents/ecosoc/docs/2000/e2000-29.htm. Viewed on 5th January 2006.

Das, Amarendra (2009). 'Regulatory Authority over Minerals: A Case for Review', *Economic and Political Weekly,* vol. 44(10), pp. 105-109.

Dennison, E. F., (1967). 'Why growth rates differ: Post war experience in nine western countries', Brooking Institution, Washington DC.

Dey, Dipankar (2001). 'Globalisation and the Indian Petroleum Industry'. Asia-Europe Dialogue. http://www.indiaresource.org/issues/energycc/2003/globpetroleumindust.html. Retrieved on 31st August 2008.

The Dharitri (2008). Online Odia newspaper. http://www.dharitri.com/160408/story13.asp. Retrieved on 06th July 2008.

Doornik, Juegen A. et al (2001). *Econometric Modelling Using PcGive 10*, vol. III., London: Timberlake Consultants Ltd.

Doornik, Jurgen A. (2008). http://www.doornik.com/products.html#PcGive.

Downing, Theodore E. (2002). 'Avoiding New Poverty: Mining-Induced Displacement and Resettlement'. Mining, Minerals and Sustainable Development. International Institute for Environment and Development.

Faria, Ricardo Coelho,; Geraldo da Silva Souza and Tito Belchier Moreira (2005). 'Public versus Private Water Utilities: Empirical Evidence for Brazilian Companies', *Economic Bulletin,* vol. 8(2), pp. 1-7.

Federation of Indian Mineral Industries (2009). 'Corporate Social Responsibility', online available at http://www.fedmin.com/html/csrdivision.htm, Retrieved on 14th November 2009.

Fernandes, W and V. Paranjpye (1997). 'Hundred years of involuntary displacement in India'. Rehabilitation Policy and law in India: A Right to livelihood, Indian Social Institute, New Delhi.

Fernandes, Ana M. (2007). 'Trade policy, trade volumes and plant-level productivity in Colombian manufacturing industries', *Journal of International Economics*, 71, pp. 52-71.

Fernades, A. M. (2007). 'Trade policy, trade volumes and plant-level productivity in Colombian manufacturing industries', *Journal of International Economics*, 71, 5271.

Frank, Windmeijer, (2000). 'A finite sample correction for the variance of linear two-step GMM estimators', IFS Working Papers W00/19, Institute for Fiscal Studies.

—. (2006). 'GMM for panel count data models', Bristol Economics Discussion Papers 06/591, Department of Economics, University of Bristol, UK.

Friedman, Milton (1970). 'The social responsibility of business is to increase its profits'. *The New York Times Magazine*, September 13.

George, A. S. (2005). 'Laws related to mining in Jharkhand', *Economic and Political Weekly*, vol. 40(41), pp. 4455-4458.

Ghazali, N. A. M. (2007). 'Ownership structure and corporate social responsibility disclosure: some Malaysian evidence', *Corporate Governance*, vol. 7(3), pp. 251–266.

Ghose, M. K. (2003). 'Promoting cleaner production in the Indian small-scale mining industry', *Journal of Cleaner Production,* Vol. 11, pp. 167–174.

Gordon, H S (1954). 'The economic theory of a common property resource: the fishery', *Journal of Political Economy,* 62, pp.124-142.

Government of India (2005). 'The Expert Committee on Road Map for Coal Sector Reforms', Ministry of Coal, New Delhi.

—. (2006). 'National Mineral Policy: Report of the High Level Committee', Planning Commission, New Delhi.

—. (2007). National Accounts Statistics-Sources & Methods, Central Statistical Organisation, New Delhi.

—. (2010). 'National Mineral Scenario'. Department of Mines, Government of India. New Delhi. http://mines.nic.in/index.aspx?lid=73&level=2&chk=24dfe45y5edf5e3 Viewed on December 19, 2010.

—. (Volumes from 1995 to 2006). 'Statistics of Mines in India'. Directorate General of Mines Safety, Ministry of Labour & Employment.

Griliches, Zvi and Jacques Mairesse (1998). 'Production Functions: The Search for Identification', in: Zvi Griliches (ed.) *Practicing Econometrics, Essays in Methods and Applications*, Edward Elgar, Cheltenham.

H. E. Frech, III (1980). 'Property rights, the Theory of the Firm, and Competitive Markets for Top Decision-Makers', *Research in Law and Economics*, vol. 2, 49.

Hardin, Garret (1968). 'The Tragedy of the Commons', *Science,* 162: 1243-48.

Hart, Oliver D. (1983). 'The market mechanism as an incentive scheme', *Bell Journal of Economics,* 14, pp.366-382.

Herbert, T., and K. Lahiri-Dutt (2004). 'Coal sector loans and displacement of indigenous populations: lessons from Jharkhand', *Economic and Political Weekly*, vol. 39(23), pp. 2403–2409.

Hesterberg, T., Shaun, Monaghan, David, S. Moore, Ashley, Clipson and Rachel, Epstein (2003). 'Bootstrap methods and permutation tests companion', Chapter 18 to the practice of business statistics, New York: W. H. Freeman and Company.

Hilson, Gavin (2000). 'Pollution prevention and cleaner production in the mining industry: an analysis of current issues', *Journal of Cleaner Production,* vol. 8. pp. 119–126.

Hodges, Caroll Ann (1995). 'Mineral resources, environmental issues, and land use'. *Science*, New Series. vol. 268(5215), pp.130555-1312.

Horowitz, Joel L. (2001). 'The Bootstrap', In: James J. Heckman and Edward Leamer (ed.) *Handbook of Econometrics Vol. 5*, Elsevier, Amsterdam: 3159-3228.

Hotelling, H. (1931). 'The Economics of Exhaustible Resources', *Journal of Political Economy*, 39(2), 137-175.

India Together (2008). 'CAG: Compensatory Afforestation a hoax in M.P'. 30 April. http://www.indiatogether.org/2008/apr/gov-afforest.htm. Retrieved on July 06 2008.

India Together (2010). 'Lokayukta slams mining in Karnataka's forests'. http://www.indiatogether.org/2010/jan/env-kamines.htm. 17 January. Viewed on 21st February 2010.

Isaac, Ehrlich; Georges Gallais-Hamonno; Zhiqiang Liu; Randall Lutter (1994). 'Productivity Growth and Firm Ownership: An Analytical and Empirical Investigation'. *Journal of Political Economy*, vol. 102(5), 1006-1038.

Jalan, Karna (2006). 'S.W.O.T Analysis of Indian Mining Industry', http://www.indianmba.com/Occasional_Papers/OP126/op126.html. ICFAI Business School. September 30. Viewed on 21st November 2011.

Jensen, J., McGuckin, R., and K. Stiroh. (2001). 'The impact of vintage and age on productivity: evidence from U.S. manufacturing plants', *Review of Economics and Statistics*, 83, pp. 323-332.

Jha, Rupa (2006). 'India –struggling over adivasi land'. Give me land, British Broadcasting Corporation, London, www.bbc.co.uk/world-serivce/worldagenda/pdf/give_me_land.pdf, as viewed on October 12, 2006.

Joskow, Paul L. (1985). 'Vertical integration and long term contracts: the case of coal burning electric generating plants', *Journal of Law, Economics and Organisation,* Spring, 1, pp. 33-80.

—. (1987). 'Contract duration and relationship-specific investments: empirical evidence from coal markets', The *American Economic Review*. Vol.77 (1), pp. 168-185.

Kalshian, R. and Kalshian, R., ed. (2007). *Caterpillar and the Mahua Flower: Tremors in India's mining fields*, New Delhi: Panos South Asia.

Karyn, Keenan, José de Echave, and Ken Traynor. (2002). 'Mining and Communities: Poverty Amidst Wealth', Political Economy Research Institute, University of Massachusetts,Conference Paper Series No.3.

Kapelus, P. (2002). 'Mining, Corporate Social Responsibility and the Community, The Case of Rio Tinto, Richards Bay Minerals and the Mbonambi,' *Journal of Business Ethics*, 39(3), pp. 275–296.

Keenan, Karyn, José de Echave, and Ken Traynor (2002). 'Mining and Communities: Poverty Amidst Wealth'. Political Economy Research Institute, University of Massachusetts, Conference Paper Series No.3.

Klein, Lawrence R. (1988). 'Components of competitiveness', *Science*, New Series, vol. 241(4863), pp. 308-313.

Klein, Benjamin; Crawford, Robert and Alchian, Armen (1978). 'Vertical integration, appropriable rents and competitive contracting process', *Journal of Law and Economics,* 21, pp. 297-326.

Krishnamurthy, K.V. (2004). 'Environmental Impacts of Coal Mining in India', in (Indra N. Sinha, Mrinal K. Ghose and Gurdeep Singh), Proceedings of the National Seminar on Environmental Engineering with special emphasis on Mining Environment, NSEEME-2004, pp. 19-20.

Kumbhakar, Subal C. and Ana Lozano-Viva (2004). 'Deregulation and Productivity: The case of Spanish Banks' Documento de trabajo, E2004/24. CentrA. Fundacion Centro de Estudios Andaluces.

Lal, V. B.; Arbol, D. K.; Bose, P. R., and Kishore, Kumar (1988). 'Environmental Impact of the Aluminum Industry' in (S.C. Joshi, G. Bhattacharya, Y. P. S Pangtey, D. R. Joshi and D. D. Dani, eds.),

Mining and Environment in India, Nainital: Himalayan Research Group, H. R. Publishers.

Lahiri-Dutt, Kuntala (2007). 'Illegal coal mining in eastern India: rethinking legitimacy and limits of justice', *Economic & Political Weekly*, vol. 42 (49), pp 57-66.

Lederman, D. and W. F. Maloney (2008). 'In Search of the Missing Resource Curse'(4766), Technical report, The World Bank, Development Research Group, Trade Team and Latin America and the Caribbean Region, Office of the Chief Economist.

Levinsohn, James and Amil Petrin (2003). 'Estimating production functions using inputs to control for unobservables', *Review of Economic Studies*, vol.70, pp. 317-341.

Li Shaomin and Jun Xia (2008). 'The roles and performance of state firms and non-state firms in China's economic transition', *World Development,* vol. 36(1), pp. 39-54.

Loayza Ismael Fernando (1999). 'Competitiveness, environmental performance, and technical change: a case study of the Bolivian mining industry' in (Alyson Warhurst, ed.), *Mining and the Environment: Case Studies from the Americas,* Canada: IDRC.

Long, K. R. (1995). 'Economics of Mining Law', *Nonrenewable Resources,* vol. 4, pp. 74-83.

Louism De Alessi (1980). 'The economics of property rights: a review of the evidence', *Rsearch in Law and Economics*, vol. 2(1), pp. 27-28.

Majumdar Sumit K (1998). 'Assessing comparative efficiency of the state-owned mixed and private sectors in Indian industry', *Public Choice.* vol. 96(1-2).

Marshak J. and Andrews W. (1944). 'Random simultaneous equations and the theory of production', *Econometrica*, 12, pp. 143-205.

Martin, S. (1993). 'Endogenous firm efficiency in a Cournot principal-agent model', *Journal of Economic Theory,* 59, pp. 445–450.

McMahon Gary, Jose Luis Evia, Alberto Pasco Font and Jose Miguel Sanchez (1999). 'An Environmental Study of Artisanal, Small, and Medium Mining in Bolivia, Chile, and Peru'.

McMahon, Gary, Ella Subdibjo, Jean Aden, Aziz Bouzaher, Giovanna Dore, and Ramani Kunanayagam (2000). 'Mining and the Environment in Indonesia: Long-term Trends and Repercussions of the Asian Economic Crisis.' An EASES Discussion paper series 21438, November.

Miller, E. (1973). 'Some implications of land ownership patterns for petroleum policy.', *Land Economics,* 49(4), pp. pp. 414-423.

Ministry of Micro, Small and Medium Enterprises (2007). 'Industrial Policy Resolution'. Government of India. http://www.laghu-udyog.com/policies/iip.htm#Indus2; retrieved on 15th April 2007.

Miwa, Yoshiro and, J. Mark Ramseyer (2000). 'Rethinking relationship-specific Investments: subcontracting in the Japanese automobile industry'. Discussion Paper No. 282. Harvard Law School. Cambridge. http://www.law.harvard.edu/programs/olin_center/.

Moody, Roger (2007). 'The base alchemist' in *Caterpillar and the Mahua Flower: Tremors', in* (R. Kalshian and R. Kalshian, ed.), *India's mining fields*, New Delhi: Panos South Asia.

Mooney Christopher Z. and Robert D. Duval (1993). *Bootstrapping A Nonparametric Approach to Statistical Inference,* New Delhi: Sage University Papers, Sage Publications.

Nickell, S. (1999). 'Product markets and labour markets', *Labour Economics*, 6: 1-20.

Nickell, Stephen J. (1996). 'Competition and corporate performance', *Journal of Political Economy*, vol. 104(4), pp. 724-46.

Nishimizu, M. and J.M. Jr. Page (1986). 'Productivity change and dynamic comparative advantage', *Review of Economics and Statistics*, vol. 68 (2), pp. 241-247.

Núñez-Barriga, A. and Castañeda-Hurtado, I. (1999). Environmental Management in a Heterogeneous Mining Industry: The Case of Peru. Chapter 41: Mining and the Environment: Case Studies from the Americas. International Development Research Centre (IDRC). http://www.idrc.ca/books/focus/828/ chapter4.html (Retrieved: May, 2009).

Page, Jr. (1986). 'Productivity change and dynamic comparative advantage', *Review of Economics and Statistics*, vol. 68(2), pp. 241-247.

O'Connor, D. and Turnham, D. (1991). 'Environmental management in developing countries: An overview', *Development and International Co-operation*, vol. 3(13), pp. 75–100.

O'Connor, D.C. (1991a). 'Market based incentives', in Environmental degradation from mining and mineral processing in developing countries: corporate responses and national policies, in (A. Warhurst, ed.) Science Policy Research Unit, University of Sussex, Brighton, Sussex, UK. Discussion document, Section 2, Ch. 5.

OECD (2001). 'Measuring Capital: Measurement of Capital Stocks, Consumption of Fixed Capital and Capital Services', OECD, Head of Publications Service, OECD Publications Service, 2, rue André-Pascal,75775 Paris Cedex 16, France. http://www.sourceoecd.org/

—. (2001a). 'Measuring Productivity: OECD Manual: Measurement of

Aggregate and Industry-Level Productivity Growth', OECD.

Office of Fair Trading (2007). 'Productivity and Competition: An OFT perspective on the productivity debate'. http://www.oft.gov.uk/shared_oft/economic_research/oft887.pdf. (Retrieved: 18th May 2008).

Olley, G. Steven and Ariel Pakes. (1996). 'The dynamics of productivity in the telecommunication equipment industry', *Econometrica*, vol. 64 (6), pp. 1263-97.

Olson, Mancur (1965). *The Logic of Collective Action*. Cambridge MA: Harvard University Press.

Ostrom, Elinor (1990). *Governing the Commons: The Evolution of Institutions for Collective Action*. Cambridge University Press.

Panda, Ranjan K. (2007). 'Under a Black Sky'. in *Caterpillar and the Mahua Flower: Tremors in India's mining fields*, in (R. Kalshian and R. Kalshian, ed.), New Delhi: Panos South Asia.

Pakes, A. (1996). 'Dynamic Structural Models, Problems and Prospects: Mixed Continuous Discrete Controls and Market Interaction,, in (C. Sims, ed.), *Advances in Econometrics, Sixth World Congress*, vol. II, Cambridge University Press, pp. 171-259.

Parameswaran, M. (2004). 'Economic Reforms, Technical Change and Efficiency Change: Firm level Evidence from Capital Goods in India'. *Indian Economic Review*, vol 39(1): 239-260.

—. (2007). 'International trade, R&D spillovers and productivity: evidence from Indian manufacturing industry'. Working Paper No. 385. Centre for Development Studies, Trivandrum, India.

Pargal, S. and Wheeler, D. (1996). 'Informal Regulation of Industrial Pollution in Developing Countries: Evidence from Indonesia', *Journal of Political Economy*, vol. 106 (6), pp. 1314-1327.

Rasmussen, R.O. and N.E. Koroleva, (2003). 'Social and Environmental Impacts in the North' as cited in http://www.freewebs.com/epgOrissa/Mine%20over%20Matter.pdf. Viewed on Februay 14, 2010.

Rangarajan, L. N. (1992). *'The Arthashastra'*, New Delhi: Penguine Books Ltd.

Sachs, J. D. and A. M. Warner (1995), 'Natural Resource Abundance and Economic Growth.'(5398), Technical report, National Bureau of Economic Research.

Sarkar, A. (2007). 'Development and Displacement Land Acquisition in West Bengal', *Economic and Political Weekly*, vol. 42(16), pp. 1435–1442.

Saxena, N.C. Gurdeep, Singh and Rekha, Ghosh (2002). 'Water Management and Environmental Management in Mining Areas', (Saxena et al., ed) Jodhpur: Science Publications, , pp 205-247.

Scharfstein, S. David (1988). 'Product market competition and market slack'. *Rand Journal of Economics,* 19, pp. 147-55.

Schmid, A. and Robin, D. K. (1995). 'Focus on environment: brazil emerges as Latin American's new environmental hotspot'. *ENR*, 234(21), pp. 42-43.

Schmidt, K.M. (1996). 'Managerial incentives and product market competition'. Centre for Economic Policy Research, DP No. 1382, April, London.

Schmitz, Jr James A. (2005). 'What determines productivity? Lessons from the dramatic recovery of the U.S. and Canadian iron ore industries following their early 1980s crisis', *Journal of Political Economy*, vol.113(3), pp. 582-625.

Sheshinski E. and L.F. López-Calva (2003). 'Privatization and Its Benefits: Theory and Evidence', *CESifo Economic Studies*, vol. 49(3), pp. 429–459, Available at http://cesifo.oxfordjournals.org/, Accessed June 17, 2006.

Smith, Jerry (2004). 'Productivity Trends in the Gold Mining Industry in Canada' Centre for the Study of Living Standards. Research Report 2004-08. Ottawa, Ontario.

Solow, R. M. (1974). 'The economics of resources or the resources of economics', *The American Economic Review,* 64(2), 1-14.

Sosa. Irene (2000). 'Mining and communities: An annotated bibliography', Prepared for Mining Watch Canada as a supplement to the workshop "On the Ground Research".

Srivastav, S. (2006). 'Environmental and social challenges of mineral-based growth in Orissa'. World Bank, New Delhi.

Srivastava, Vivek (1996). *Liberalisation, Productivity and Competition: A Panel Study of Indian Manufacturing*, New Delhi: Oxford University Press.

STATACORP (2007). *STATA base Reference Manual*: vol. 2(I-P), Release 10.STATACorp L P, Texas.

Stiglitz, Joseph. E. (1976). 'Monopoly and the Rate of Extraction of Exhaustible Resources', *American Economic Review*, vol. 66(4), pp. 655-661.

Stiglitz, Joseph E. (1988). *Economics of Public Sector,* W.W.Norton, New York.

Symeonidis, George (2007). 'The Effect of Competition on Wages and Productivity: Evidence from the UK', Discussion Paper Series No. 626 March, *University of Essex.*

Szablowski, D. (2002). 'Mining, Displacement and the World Bank: A Case Analysis of Compania Minera Antamina's Operations in Peru', *Journal of Business Ethics*, x, vol. 39(3), pp. 247–273.

Tang, Meng-Chi (2009). 'Contract Structure and Relationship-Specific Investments: Empirical Evidence from Coal Markets', Paper presented at Taiwanese Economic Association Annual Meeting. National Chung Cheng University, Min-Hsiung, Chiayi, Taiwan.

Third World Network (1997). 'Earth Summit Plus 5 Briefing No. 5: Mining Activities (Excluded in Agenda 21) Causing Social and Ecological Problems'. http://www.twnside.org.sg/title/brie5-cn.htm. Viewed on 3rd July 2010.

Tilton, J.E., Landsberg, H.H. (1999). 'Innovation, productivity growth, and the survival of the U.S. copper industry', in David Simpson, R. (ed.), *Sources of Productivity Improvement in U.S. Natural Resources Industries*, Resources for the Future, Washington, DC.

United Nations Economic Commission for Europe Statistical Division (2003). 'Measurement of Capital Stock In Transition Economies' Conference Of European Statisticians. Occasional Paper 2003/1.

Valdiya, K. S., (1988). 'Environmental Impacts of Mining Activities' in (S. C. Joshi, G. Bhattacharya, Y. P. S. Pangtey, D. R. Joshi, and D. D. Dani, eds.), *Mining and Environment in India*, Nainital: Himalayan Research Group, H. R. Publishers.

Wang, Hua (2000). 'Pollution Charges, Community Pressure and Abatement Cost of Industrial Pollution in China', Policy Research Working Paper No. 2337. World Bank. Washington DC.

Wang, Hua and Yanhong Jin (2002). 'Industrial Ownership and Environmental Performance: Evidence from China', World Bank Policy Research Working paper 2936, December, Washington DC.

Warhurst, Alyson (1999). 'Environmental Regulation, Innovation, and Sustainable Development', in (A. Warhurst, ed.), *Mining and the Environment: Case Studies from the Americas'*, Canada: International Development Research Centre.

Warhurst, Alyson C and G. Bridge (1997). 'Economic liberalisation, innovation, and technology transfer: opportunities for cleaner production in the minerals industry', *Natural Resources Forum*, vol. 21(1), pp. 1-12.

Williamson, Oliver E. (1979). 'Transaction-cost economics: the governance of contractual relations'. *Journal of Law and Economics,* 22, pp. 233-61.

—. (1983). 'Credible commitments: using hostages to support exchange', *American Economic Review*, 73, pp. 519-40.

—. (1985). *The Economic Institutions of Capitalism*, New York: Free Press.

Windmeijer, F. (2000). 'A finite sample corrections for the variance of linear two-step GMM estimators', IFS working paper W00/19 (www.ifs.org.uk), London: The Institute for Fiscal Studies.

The World Bank (1993). 'Types of development projects causing displacement'. http://www.forcedmigration.org/guides/fmo022/fmo022-3.htm. Viewed on 5/25/2010. World Bank Technical Paper No. 429. The World Bank. Washington D.C.

Ying John S. (1990). 'The Inefficiency of Regulating a Competitive Industry: Productivity Gains in Trucking Following Reform', *Review of Economics and Statistics,* vol. 72(2), pp. 191-201.

NAME INDEX

SUBJECT INDEX